Monographs on soil survey

General Editors
P. H. T. BECKETT
V. C. ROBERTSON
R. WEBSTER

Soil and permafrost surveys in the arctic

KENNETH A. LINELL

and

JOHN C. F. TEDROW

CLARENDON PRESS · OXFORD
1981

Oxford University Press, Walton Street, Oxford OX2 6DP

OXFORD LONDON GLASGOW
NEW YORK TORONTO MELBOURNE WELLINGTON
KUALA LUMPUR SINGAPORE JAKARTA HONG KONG TOKYO
DELHI BOMBAY CALCUTTA MADRAS KARACHI
NAIROBI DAR ES SALAAM CAPE TOWN

Published in the United States by Oxford University Press,
New York

British Library Cataloguing in Publication Data
Linell, Kenneth A
Soil and permafrost surveys in the arctic.—
(Monographs on soil survey).
1. Soil-surveys—Arctic regions
2. Frozen ground—Arctic regions
I. Title II. Tedrow, John C F III. Series
631.4′998 S599.9.A/ 80-41844

ISBN 0-19-857557-2

Printed in the United States of America

Preface

The problems of arctic soil and permafrost surveys are particularly important because arctic lands have a number of characteristics usually not found in other parts of the globe: widespread distribution of permafrost, unusual terrain conditions, and extreme climate, among others.

For centuries there was little interest in the arctic and accordingly there was little apparent need for accumulating data. Since the Second World War, however, there has been a dramatic change in the picture. There is now need for modern science and technology in site selection and in the construction and maintenance of airfields, roadways, pipeline corridors, buildings, and related structures. Further, in considering such questions as the northern extension of agriculture, land use, conservation, and wildlife, soil and permafrost conditions become critical. Most books on soil mechanics and pedology give little space to frozen soil conditions. Our aim in this volume is to focus on the special problems associated with surveying arctic soils and permafrost. In so doing we have not attempted to cover the arctic literature completely. This already has been done in a number of other volumes. We have focused our discussions on field problems and procedures together with a multiplicity of unique situations associated with arctic terrain.

Originally we had considered using the metric system exclusively throughout the text, as had been done with other volumes in this series. The nature of the subject, however, convinced us that it was inadvisable to carry out the original plan completely. Many of the publications we cited had already used Imperial or (American) Customary Units of Measurement, and converting these units would in some instances have created a number of problems. Where appropriate, however, we have used the metric or dual system of measurement. Furthermore, we have retained units of measurement as originally stated in publications. These units are generally followed by alternate units. To help the reader convert certain measurements, we have appended a list of conversion factors, together with a list of certain abbreviations, near the end of the volume.

In compiling this volume we acknowledge the counsel of Dr. Jerry Brown of the Cold Regions Research and Engineering Laboratory, Hanover, New Hampshire. The invaluable assistance of Mr. Wesley

Pietkiewicz and Mr. Harold Larsen of the same laboratory during assembly of the illustrative material is also gratefully acknowledged.

Hanover, New Hampshire KAL
New Brunswick, New Jersey JCFT
August 1979

Contents

Monographs on soil survey

1. Delineation and structure of the arctic

The arctic defined

In a general sense nearly everyone would probably agree what is meant by the term *arctic,* but when one attempts to define it with precision or to plot specific boundaries on a map, the problem becomes rather complex. Most naturalists prefer to use the vegetative concept in which land north of the tree line is considered arctic. Such a definition is not without limitations, however, because there are circumpolar as well as latitudinal changes in floristics, migration of plants, and development of plant communities plus additional factors, such as the presence of permafrost, recently deglaciated landscapes, and still other related factors which need to be considered. Further, the arctic tree line is interrupted by mountain ranges, ice fields, and large bodies of water. In reality the northern margin of the forest does not everywhere form a clearly defined, continuous line. Instead it is more a blend or mosaic of forest and tundra elements. Hustich (1966) describes the mosaic as a forest–tundra transition zone which is composed of (i) timberline, (ii) physiognomic forest line, (iii) tree line, (iv) tree species line, and (v) historical tree line. Despite some limitations of a vegetative definition, use of the tree line as the southern margin of the arctic is realistic for pedologic purposes. Such a definition, as used earlier by Polunin (1951), represents a compromise but yields a generalized dividing line. Insofar as zonation of arctic lands for pedologic purposes is concerned, we base our discussions on that information shown in Fig. 1.1. The tree line approximates the Arctic Summer Frontal Zone (Bryson and Hare 1974), so indirectly we are using in part an index which primarily reflects summer climate.

We now turn to the problem of defining and delineating the arctic region for engineering purposes. One may casually assume that arctic is a clearly defined term acceptable in all disciplines, but such is not the case. Whereas naturalists tend to focus on biota in recognizing zones and making delineations, such a procedure in reality is an indirect method of recording climate and other related parameters. In defining the arctic for engineering purposes the main focus has been on the use of ambient temperatures. Obviously this approach has great practical significance. Engineers generally consider the arctic to be 'The northern region in which the mean temperature for the warmest month is less than

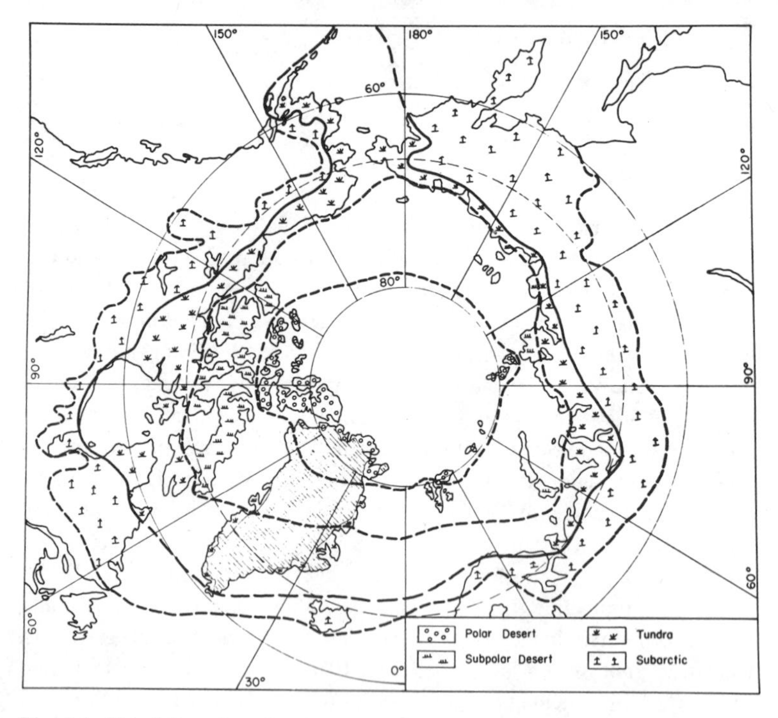

Fig. 1.1. The delineation of major arctic zones on a basis of vegetation.

50°F and the mean annual temperature is below 32°F' (Stearns 1965). The area so delineated is shown in Fig. 1.2.

Other criteria have been used to define the arctic. The *Nordenskjöld line* had been proposed to mark the southern edge of the arctic. The line is established by using the formula $V = 9 - 0.1\,K$, where V is the mean temperature of the warmest month and K is the mean temperature of the coldest month.

The Arctic Circle as well as other parallels of latitude have been considered in defining the arctic, but field studies of the north will show that such a concept must be largely abandoned.

Nearly all arctic lands have permafrost, but permafrost also extends well south of the southern margin of the arctic. This condition is well exemplified in Canada but is even more striking in central and eastern Siberia, where the southern margin of the arctic zone usually occurs at latitudes south of 70°N but the southern margin of the permafrost zone extends south of 50°N in parts of eastern Asia.

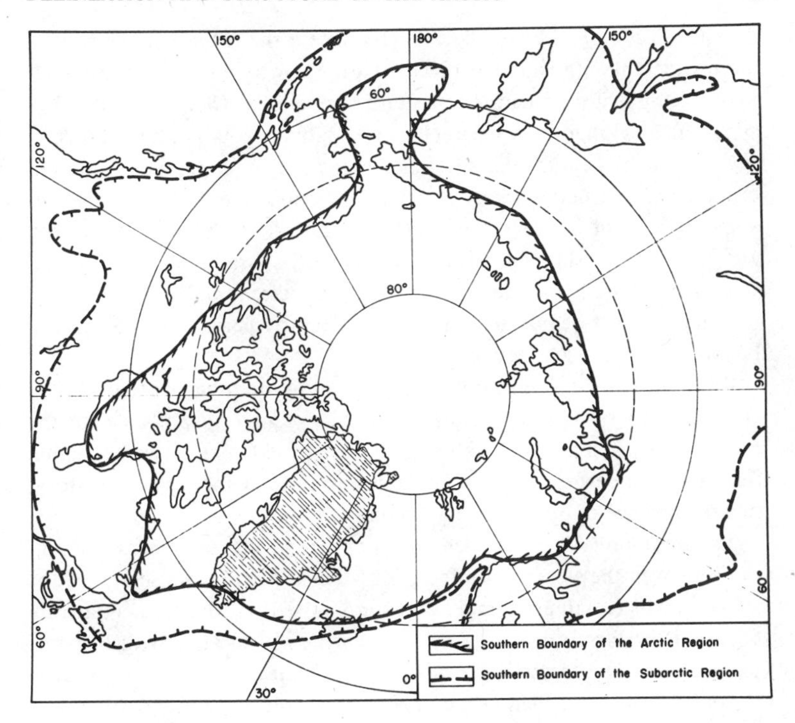

Fig. 1.2. Arctic areas defined according to temperature (Stearns 1965).

Between the boreal forests with a closed canopy and the tree line is a wide expanse of land naturalists refer to as 'taiga' (land of the little trees), 'pretundra,' 'forest tundra,' 'hemiarctic,' or simply 'subarctic' (*Ecology of the Subarctic Regions* 1970), which combines both arctic and forest elements. The term *subarctic* appears to be easily definable, but defining it in detail becomes quite complex. Using the above definition, however, we can generalize a subarctic zone (Fig. 1.1). The southern boundary of the subarctic, particularly in Siberia, becomes strikingly indefinite, especially where it becomes influenced by and confluent with the alpine zones. In western Alaska arctic and subarctic zonation also becomes rather difficult to define.

As is the case with depicting an arctic (tundra) zone, the subarctic defined by engineers is not entirely the same as that proposed by naturalists. Because of the overwhelming influence of ambient and soil temperatures on the properties and function of soil, soils engineers usually define the subarctic as "The region in which the mean temperature for the coldest month is below 0°C (32°F), where the mean temperature of the warmest month is above 10°C (50°F), but where there are less than four months with a mean temperature above 10°C (Stearns 1965). The subarctic as plotted by Stearns (Fig. 1.2) extends south of that shown in Fig. 1.1, the reason for the difference being the employment of different criteria in constructing the two maps. It is realistic to state that there is some agreement between the southern limit of the subarctic (Fig. 1.2) and the southern limits of discontinuous and sporadic permafrost, but the two lines are by no means coincident on a circumpolar basis.

Whereas most of the subarctic is underlain by permafrost, it tends to be covered by a thick, active soil layer.

Returning to the delineation of the subarctic on a basis of vegetation (Fig. 1.1), there may be factors other than climate responsible for the plant cover, such as relief features, dearth of soil material (well exemplified along the eastern coast of Hudson Bay), fire history, glacial history, and occurrence of open muskeg, among others.

The term *high arctic* has been used by physical geographers and naturalists over the years. In effect, it is the northernmost sector of the arctic in which summer temperatures are quite low, climate is relatively dry, and the terrain has a stony, raw earthy appearance with a sparse plant cover. The terms *polar desert* and *arctic desert* have been used synonymously with high arctic. The high arctic has a more severe climatic environment than the main tundra belt, a fact attested to by natural plant and animal distribution, as well as by meteorological observations. In general the 40°F July isotherm can be used as a criterion for separating the high arctic (polar desert) from the main arctic. Permanent Eskimo settlements have not penetrated far into the high arctic.

The southern margin of the high arctic (polar desert) blends with the warmer, wetter tundra landscape. This transition zone is shown in Fig. 1.1 and approximates the term *subpolar desert, arctic tundra,* or *midarctic*.

Geology and physiography

Two prominent, ancient, crystalline rock masses extend well into the arctic zone: the Canadian Shield, which extends from eastern Canada westward to the vicinity of Great Bear Lake and Great Slave Lake; and the Baltic Shield, which includes northern Fennoscandia and the Kola Peninsula (Fig. 1.3).

From the White Sea eastward to the Ural Mountains the East Euro-

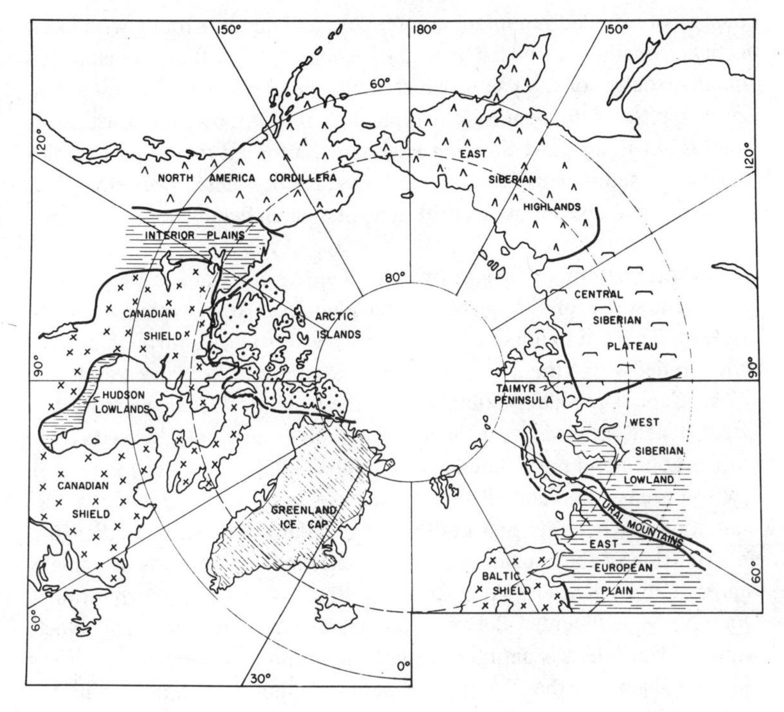

Fig. 1.3. Geologic and physiographic regions of the arctic.

pean Plain is made up of sedimentary rock with a thick cover of glacial drift. In the northern sectors, particularly the Kanin Peninsula, the surficial mantle has some marine facies. The Ural Mountain complex, including Novaya Zemblya, is a fold zone of Paleozoic age bedrock. The glacial drift cover is thin with prominent rocky conditions. East of the Ural Mountains to the Yenisei River is the West Siberian Lowland, made up of till-covered sedimentary rocks. The northern portion of the province, as exemplified by Yamal, also has low relief features. Within the West Siberian Lowland much of the land is poorly drained with extensive muskeg formation (Sjors 1961). The Central Siberian Plateau occupies the land between the Yenisei and Lena Rivers. This region is represented by the Anabar Shield and peripheral sedimentary rocks resting on the Siberian Platform. That portion of the plateau within the arctic and subarctic is till-covered. To the north, conditions of the Taimyr Peninsula are similar to those of the northern Urals in that the area consists of a folded mountain complex of sedimentary rocks with some vertical zonality of soils. On a macroscopic scale, the peninsula has shallow, rocky soils, some of which have been referred to as *polar desert*. The base of the peninsula is typical of tundra, however. East of the Lena River is the East Siberian Highlands, a folded mountain complex made up of sedimentary rocks including volcanics. The lower elevations have local and discontinuous drift deposits which become more conspicuous to the north.

The northern rim of Siberia from the Taimyr eastward to the Kolyma River is a series of low plateaus and plains, most of which are drift-covered. The main course of the Lena River also flows through an area of low relief with a thick surficial mantle (Sachs and Strelkov 1961).

The Canadian Shield extends from the Atlantic Ocean westward to Great Bear Lake and Great Slave Lake. This is a complex area covered with stratified and unstratified glacial deposits but it includes many shallow soil areas, as exemplified by conditions in northern Quebec. Baffin Island, on one hand, is a part of the Arctic Islands but, on the other, it is an extension of the Canadian Shield. The Hudson Lowlands are chiefly represented by Lower Paleozoic rock with a covering of poorly drained Quaternary sediments. Between the Canadian Shield and the North America Cordillera is an interior platform, commonly referred to as the 'Interior Plains' or the 'Western Canadian Sedimentary Basin,' which is a series of Devonian and Cretaceous formations that generally increase in thickness westward. The western margin of Canada is occupied by the folded North America Cordillera with extensive shallow soil areas.

The bedrock of Greenland can be considered an extension of the Canadian Shield. Rocks exposed on the periphery of the Greenland Ice Sheet are mainly complex shield rocks, some of which are highly metamorphosed.

Arctic Alaska, on a megascopic scale, is an extension of conditions represented in northwest Canada. The Brooks Range is a rocky rugged east–west chain of mountains of largely Paleozoic rocks which merge to the north with the Arctic Foothills. This latter province is largely represented by Cretaceous sediments, which in turn are mantled with a discontinuous cover of glacial drift. The northern margin of Alaska consists of a coastal plain.

Returning to Canadian conditions, the Arctic Islands consist primarily of Paleozoic and Mesozoic sedimentary rocks, with some extension of the Canadian Shield being represented, especially on Baffin Island.

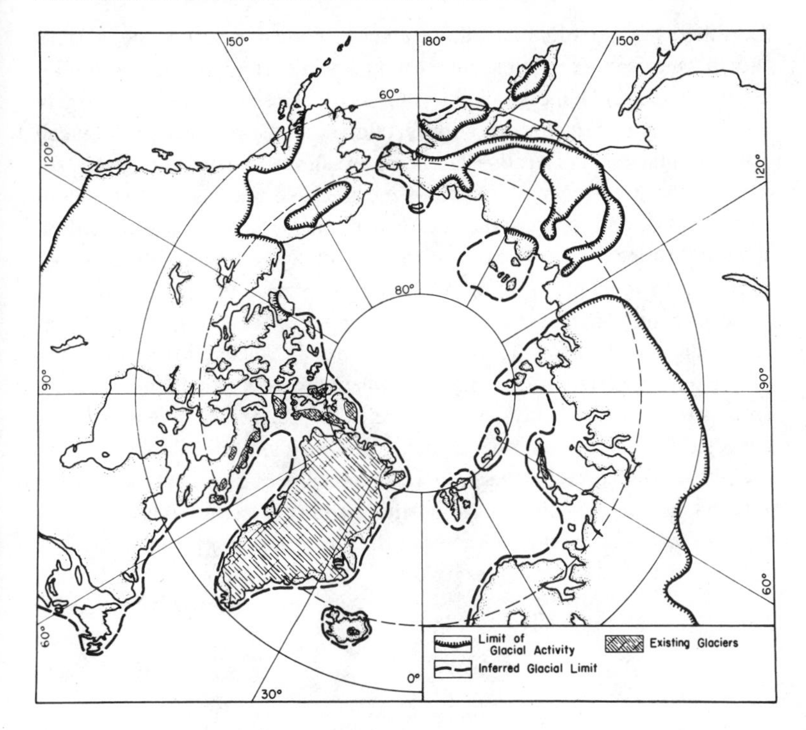

Fig. 1.4. Glacial activity in the arctic.

Glacial activity in the arctic

Nearly all arctic lands were glaciated during the Pleistocene epoch (Fig. 1.4). Glaciers largely removed existing soils, scoured and eroded bedrock, and reworked and deposited till. Other events associated with glacial activity resulted in the formation of lakes, blocking of streams, and the altering of drainage patterns, a condition favoring muskeg development.

There were at least two great centers of ice accumulation during late Pleistocene time: Baltic ice, which covered Scandinavia and extended south and east to other parts of Eurasia; and the Laurentide ice, which covered vast areas of Canada and the northern United States. Ice also accumulated in the East Siberian Highlands, the northern Rocky Mountains, the Alaskan mountains, as well as the Baffin–Ellesmere complex.

Cordilleras of mountain and alpine glaciers persist in part of the western North American mountains and particularly in eastern Canada on Baffin, Ellesmere, Axel Heiberg, and Devon Islands. Greenland remains largely ice-covered. Except for a few uncertainties in the northwestern Queen Elizabeth Islands, all Arctic islands are presumed to have been glaciated.

With the great accumulations of ice in eastern Canada, Greenland, Scandinavia, and Siberia, the land was depressed below sea level. Following disappearance of most of the ice load, isostatic adjustment occurred with major upwarping of the earth's crust. In the Hudson Bay region, rebound amounted to as much as 250 m (Lee 1962), and in northern Fennoscandia some lands rebounded nearly 900 m (Hoppe 1959).

The emerged landforms have sediments with composition ranging from wet lacustrine clays on level relief to sands and even boulders associated with various relief forms. Thus the Tyrell Sea deposits in the vicinity of Hudson Bay, the Yoldia Sea deposits of Scandinavia, and the interglacial sea deposits on the northern rim of Siberia have all had an indirect influence on the character of the soil pattern.

2. Climate and hydrology

Climate

Arctic climate is characterized by low temperatures, a small quantity of precipitation, and extended periods of daylight and darkness. Three major categories of climatic region, exclusive of the glaciers and ice fields, are shown in Fig. 2.1: polar, arctic, and subarctic. The regions, although not precise, give a general picture, particularly in reference to soil properties. The subarctic climate is further divided into two categories: subarctic (ortho) and subarctic–continental. The latter subcategory, being less humid, leaves its imprint on the soil pattern. In

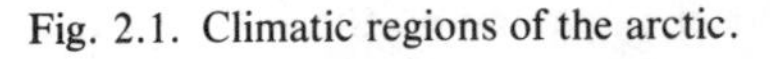
Fig. 2.1. Climatic regions of the arctic.

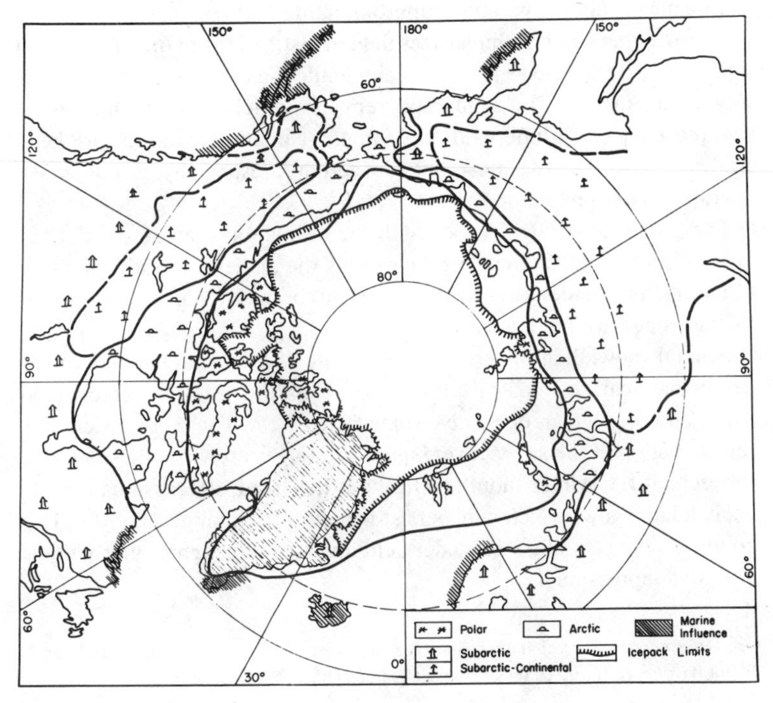

addition to the two main categories of subarctic climate, the maritime influence of certain coastal zones is also shown.

Precipitation values in the arctic are usually less than 25 to 30 cm per year decreasing to about one-half this quantity in the high arctic.† Land masses, such as those of northern Norway, southern Greenland, the coast of Labrador, and those lands bordering the Bering Strait, however, have much greater precipitation values than do the main continental areas. About one-half of the precipitation occurs as rainfall throughout much of the arctic, but in the high arctic precipitation is mostly in the form of snow. Snowfall may occur in the arctic during any month of the year, but summer snows are not lasting.

Most arctic lands have mean summer temperatures approximating 5 to 10°C for July, the months of June and August being some 3 to 5°C lower. Major elevations of land, such as those of the Ellesmere–Baffin Mountain complex and the interior of Greenland, have much lower summer temperatures than adjacent lower areas. The high arctic has summer temperatures lower than the main arctic (tundra) belt with the northern extremities of land, such as Franz Josef Land and northern Greenland, having mean July temperatures approximating 1 to 3°C.

Because this volume emphasizes field investigations of arctic soils, we point out that the field season is usually limited in most locations to June, July, and August. The landscape remains largely frozen and snow-covered until early June, and after August climatic conditions are generally unfavorable for most outside activity, especially if one must operate from a tent camp.

The period of daylight varies with the latitude as well as the time of year. With extended daylight periods plus the angle of the summer sun, potential differences in soil development between north- and south-facing slopes are minimized.

Annual snowfall in the arctic approximates 1 m, slightly higher values occurring near the coastal zones. Some of the interior sectors and especially the high arctic receive some 50 to 80 cm. There is considerable redistribution of the snow cover; spurs and ridges are virtually snow-free throughout the winter months. On the other hand, depressions tend to accumulate snow, which may persist well into the summer months. This condition occasionally will hinder field investigations, particularly in the cols and depressions.

† Temperature and precipitation values for important arctic weather stations are listed in *World Weather Records*, U.S. Government Printing Office, Washington, D.C.

Continental drainage systems

The major river systems of the arctic regions of Eurasia are quite different from those of North America. In Siberia the Ob, Yenisei, Lena, and Kolyma, among other large rivers, flow in a northerly direction, a condition which not only affects the seasonal breakup of sea ice of the northern rim of Eurasia, but also provides river access to the arctic mainland. In North America, the Mackenzie is the lone, major north-flowing river. From the Mackenzie River eastward to the Atlantic Ocean, there is little access to inland areas via navigable rivers. This condition probably played an important part in delaying development of the Kazan Region (the area between Great Bear Lake and Hudson Bay) in the Northwest Territories of Canada.

Major rivers of the arctic begin breakup in their upper reaches about mid-May, but not on the coastal fringes of the mainland until the middle to the end of June. Rivers and lakes usually freeze during October, although at higher latitudes they tend to freeze somewhat earlier. Rivers and lakes generally freeze to a depth of up to 2 m, but local measurements show great variations.

Sea ice

The Arctic Ocean is largely covered by a floating ice mass up to 10 or more meters thick, although an average value would probably be less than one-half this value. The ice is perennial but is constantly moving at a very slow rate of speed, resulting in the formation of leads, pressure ridges, and telescoping ice sheets. Figure 2.1 shows the area occupied by the permanent sea ice. In addition to the permanent ice pack, seasonal freezing of the marine waters also occurs. As a result the ice pack, at its maximum seasonal formation, occupies about twice the ocean area that it does during late summer. The seasonally formed ice plus that which breaks off from the margins of the permanent ice pack move in various directions but generally waste away at lower latitudes.

3. Permafrost and seasonally frozen ground

Permafrost is defined as the thermal condition in soil or rock of having temperatures below 0°C persist over at least two consecutive winters and the intervening summer (Brown and Kupsch 1974). Permafrost is a condition rather than a substance, contrary to common usage (Fig. 3.1). Permafrost or perennially frozen ground has long been recognized as a phenomenon of northern regions. Formation of permafrost probably started during the Pleistocene epoch or perhaps a little earlier.

Fig. 3.1. Permafrost along the Aldan River, Siberia. A 5-foot active layer overlies the perennially frozen sediments.

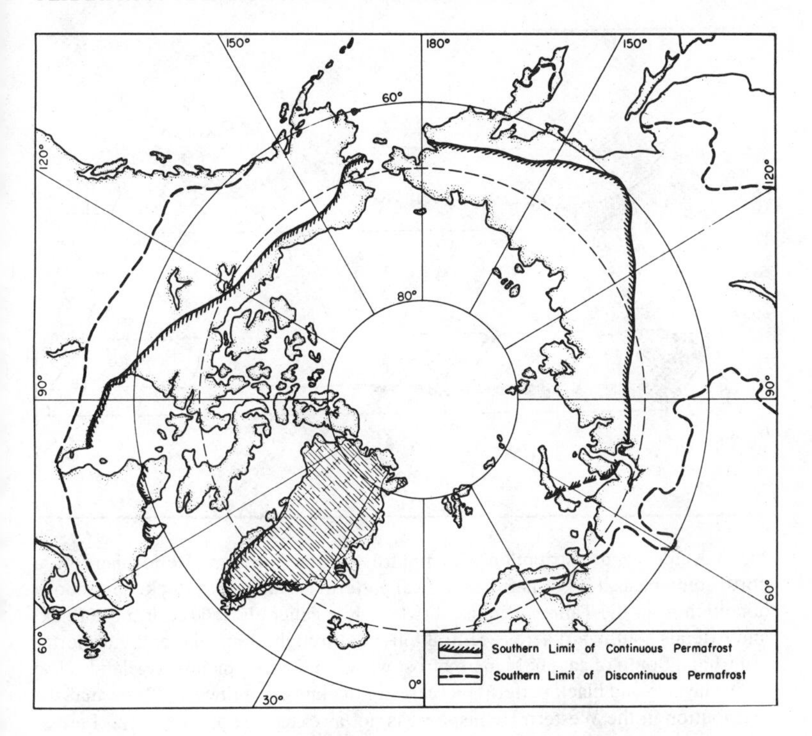

Fig. 3.2. Permafrost distribution in the Northern Hemisphere.

Distribution of permafrost

Permafrost occupies about 20 percent of the land surface of the globe. South of the major permafrost zones of the north, shown in Fig. 3.2, permafrost also exists in the high mountains (Brown and Péwé 1973). Outstanding examples are present in the Rocky Mountain Cordillera as well as the Tibetan Plateau (Baranov and Kudryavtsev 1966). The extreme northern land masses have a deep, continuously frozen mantle which, in northern Alaska, may be as much as 650 m thick. It has been measured as 400 m on Cornwallis Island, Canada. Along the Markha River of Siberia subfreezing temperatures were recorded to a depth of 1500 m, a world record (Gravé 1968). Permafrost underlies some shallow coastal waters of the Arctic Ocean, some of which is believed to be of a fossil nature; in other situations it apparently is aggrading. On a meridional basis, permafrost can be viewed as being wedge-shaped with both

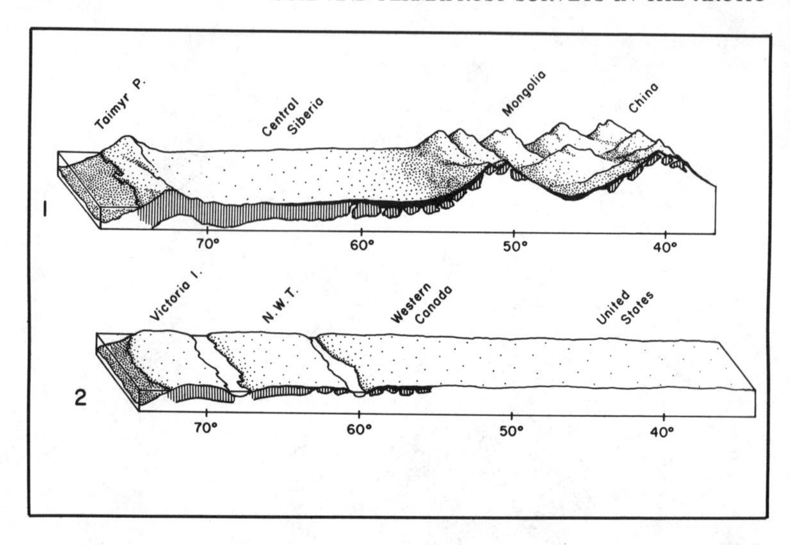

Fig. 3.3. Idealized distribution of permafrost in the Northern Hemisphere on a north-south basis. (1) Permafrost (vertical pattern) is shown as a thick, continuous condition from the Taimyr Peninsula (where it is generally 1500 ft or more thick) and extends southward across central Siberia, through Mongolia, and into northern China. South of ca. 60° N, permafrost wanes and exists mainly as islands. The active layer (solid black pattern) increases in thickness southward. (2) Permafrost distribution in the Western Hemisphere is not as extensive as in Siberia. Permafrost of the Queen Elizabeth Islands is not depicted in the diagram.

depth and thickness of the permanently frozen strata decreasing southward. Fig. 3.3 shows, in part diagrammatically, two transects—one covering central Siberia in a north–south direction and the other for central Canada also in a north–south direction. It can be seen in Fig. 3.3 that permafrost extends farther south in the Eastern Hemisphere than in the Western Hemisphere—a condition probably accounted for in part by the higher elevations of the more southerly locations in the former. Even where the permafrost table may be quite deep, especially as is the case within the discontinuous zone, it still has a regulatory influence on the behavior and performance of surface conditions.

The more southerly parts of the permafrost zone are of special interest because it is within these locations that the most vigorous northerly expansion of roadways, utility corridors, buildings, and agriculture is taking place. South of the continuous and discontinuous permafrost zones one still encounters isolated sectors of permafrost as exemplified in the higher elevations of the Rocky Mountains, Northern Scandinavia, the Altai-Sayan Range of Central Asia, and others. Isolated pockets of permafrost are also present in many of the bogs and muskegs well south of the main permafrost zone.

Nature of permafrost

Permafrost is generally envisaged as frozen sediments and ice but it also may consist of bedrock and organic material. Thus a variety of material is included under the title of permafrost, but all are united on a basis of temperature. In some situations the matrix may consist largely of ice, but in other cases permafrost may be a mixture of frozen mineral material and ice. Ice may fill voids between the mineral particles and it is common for ice lenses, veins, layers, and wedges to be prominent components. In other situations the ice content may be so low that the condition is referred to as dry permafrost.

Apart from the varied character of the mineral and organic fractions within permafrost beds, it is necessary to describe the multiplicity of other conditions which may exist. The quantity of ice in permafrost may be substantial. Soil and rock pores generally contain ice or a combination of ice and unfrozen water. Interspersed with frozen mineral material may be ice lenses and veins with complex configurations. The quantity of ice in the soil will show great variation with some examples consisting of little other than clear ice as exemplified by ice wedges. On the other hand the dry permafrost sites will generally have only small quantities of ice present. Ice wedges are widespread throughout the permafrost zone. As annual freeze–thaw cycles occur in the active layer and as growth of ice occurs, there is a massive, slow-mixing effect within the soil, a condition that is of a critical nature to soil formation as well as to engineering.

Temperature regimes

The maximum depth of seasonal thaw coincides with the top of the permafrost or, as it is commonly known, the permafrost table (Fig. 3.4), unless there is an intervening residual thaw layer. Permafrost itself even though continuously at or below 0°C, is subject to certain seasonal temperature changes. The amplitude of annual temperature change may be as much as 20°C or more near the top of the permafrost but decreases

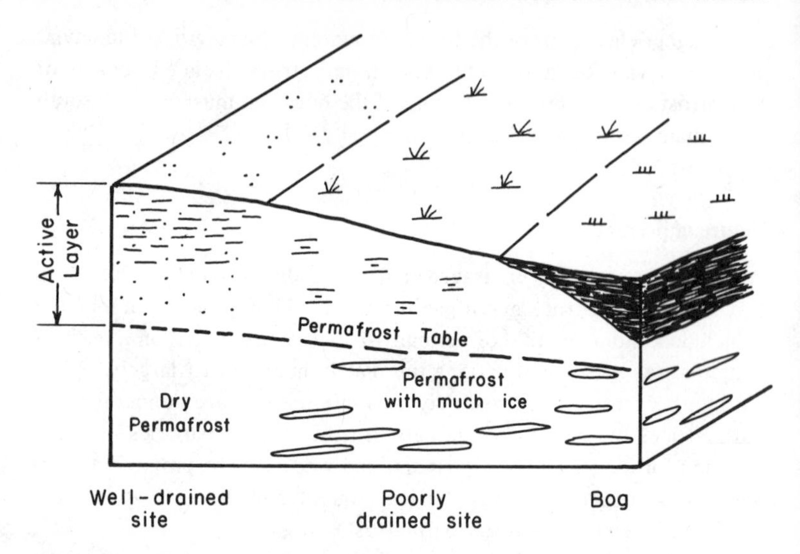

Fig. 3.4. An idealized landscape cross section showing higher, well-drained ground on the left and lower, peaty ground on the right. From left to right the permafrost table occurs nearer the surface and the permafrost has greater quantities of ground ice present.

with depth. Below a depth of some 15 to 20 m there is only a small annual change in temperature (Lachenbruch 1962).

Subsurface conditions

The greatest quantities of ground ice, including ice wedges, are usually found in level, undulating, and sloping areas where there is appreciable thickness of unconsolidated sediments. Living vegetation, the surface vegetative mats, and peat act as efficient insulators; hence, in any situation in which there is considerable organic covering on the soil, the permafrost table will occur much higher. Once the vegetative cover is destroyed, however, the permafrost degrades, resulting in a lowering of the permafrost table. There are a number of local factors, such as textural composition of the sediments, quantity of ice in the ground, and other site parameters which have a regulatory influence on the rate of permafrost degradation.

Where perennially frozen ground has only a small amount of ice present the condition is referred to as dry permafrost. The presence of ice is limited largely to the contact points between mineral grains, with pores and voids remaining air filled. Dry permafrost tends to be present in well-drained gravelly and rocky deposits such as deltaic deposits, outwash, dunes, ledges, and other freely drained sites. The initial reaction may be to focus on such sites for building because of good drainage and apparent ground stability, but experience has shown this may be only a superficial condition. As one excavates such sites the first meter or so may have relatively little ice but at greater depth large quantities of excess ice can be present. There are numerous case histories in which structural damage or failure has resulted from ground ice conditions. This situation prevails even in bedrock. In order to provide for such potential problems it is advisable to drill to obtain frozen cores and from the information gained, make necessary provisions in foundation and structural designs.

Permafrost is present throughout nearly all of the arctic with most scientific reports focusing on the so-called tundra conditions. In the high arctic (roughly that land north of 75°N) conditions are somewhat different from those of the main tundra belt in that the climate of the former being colder and drier tends to inhibit development of a complete plant cover. The drier, less-vegetated terrain gives rise to an active layer which may be a meter or more deep. In water-saturated depressions, however, maximum seasonal thaw will generally be much less, amounting to only some 20 to 30 cm.

Permafrost is present in nearly all of the treeless areas of the arctic but the perennially frozen condition also extends well into the forested areas. Vast areas of permafrost of central and eastern Siberia as well as large areas of North America are covered with woodland or forest. Factually the forest cover tends to maintain the ground ice in its frozen state.

Some of the more pressing soil problems occur within the subarctic because of poorly drained landforms on flat terrain (Radforth and Brawner 1977). Examples of these poorly drained conditions are found in the Hudson Lowlands, Northern Fennoscandia, and that large Soviet sector from the White Sea eastward to the Yenisei River among others.

Although a large body of valuable information has been accumulated on permafrost over the years, the actual thickness of the permafrost is known only in general terms. Most of the information on thickness of the permafrost has come from drill holes and extrapolation of subsurface temperature data measured from partial permafrost penetrations.

In the arctic sectors of Alaska, Canada, and Siberia, extensive drilling for engineering investigations, water supply, research purposes, and oil and gas exploration during the past thirty years or so has provided opportunity to record the depth and character of the frozen material. In addition to determination of the physical nature and condition of the surficial materials and bedrock, thermistors or other temperature-sensitive devices can be implanted in the drill hole to obtain temperature measurements. When the permafrost table is beyond the reach of hand probes, geophysical techniques may be used to approximate the depth to frozen materials. General seismic operations in cold regions have been discussed by Roethlisberger (1972). Conventional seismic studies may be used to map and determine the depth to the top of the frozen material, but are not suitable for establishing the actual permafrost thickness (Barnes 1966).

Seasonally frozen ground

The upper part of the soil which undergoes seasonal freeze–thaw cycles is termed the active layer (Fig. 3.4). In the high arctic, seasonal thaw begins in early June on the drier, exposed sites, and by the end of June about 75 percent or more of the seasonal thaw may have been reached, by late August or early September the entire mantle of soil refreezes. The top of the permafrost is designated as the permafrost table. The depth at which frozen conditions are encountered while thaw is still progressing during early summer is not the permafrost table. Instead the permafrost table coincides with the maximum depth of seasonal thaw, if there is no residual thaw layer. The receding seasonal frost table is commonly referred to by Swedish investigators as *tjaele*. Within the permafrost itself unfrozen bodies of material, known as *taliks* may occur.

In the subarctic regions the underlying permafrost may lie up to several meters or more below the ground surface, and where a degrading condition exists the seasonally frozen condition may not extend downward to the permafrost. This condition becomes more pronounced southward.

Factors determining the active layer thickness

Climate directly and indirectly is of great importance in establishing the thickness of the active layer. Frozen snow-covered areas remain frozen as long as the snow cover persists. In the high arctic, wet meadows may not begin to thaw until July, and in some depressions an unusually thick snow cover may last from one year to the next. But throughout most arctic lands, June is the month of major soil thaw with well-drained, poorly vegetated topographic positions tending to thaw earlier and deeper than the low, wet, sod-covered areas. In the subarctic, thaw may start as early as April.

During freeze-up, the situation is reversed in that the high, dry positions freeze first. But the overall freeze-up is generally compressed into a shorter period than the spring and summer thaw.

Soil texture has a pronounced regulatory influence on the rate and depth of seasonal thaw. Coarse textured, free-draining materials such as sands and gravels thaw much faster than the heavier textured moisture-retaining silts and clays. This situation is particularly in evidence on the high positions where there is generally less seasonal snow cover.

4. Thermal regime

The thermal regime in the ground is the pattern of temperature variations occurring with time and with depth from the surface. In winter, heat flows out of the ground; in summer it flows in. In the upper 30 to 60 ft (approx. 10 to 20 m) of the ground an annual pattern of temperature variations occurs, as indicated in Figs. 4.1 and 4.2, in response to the temperature changes at the ground surface. Within the upper few feet of the ground, variations occur corresponding to fluctuations of surface temperature over shorter periods, with diurnal variations occurring in the uppermost 1 or 2 ft (0.3 or 0.6 m). The temperature variations in the ground lag in time behind those occurring at the surface and decrease in amplitude with depth. In Fig. 4.2, for example, temperatures between depths of 16 and 31 ft (4.9 and 9.4 m) were at their warmest on 11 February, even though temperatures at the ground surface were near their coldest extreme, and at a depth of 31 ft (9.4 m) the amplitude of annual temperature variations was about 3°F (1.7°C). The depth below which the amplitude of the temperature variations becomes imperceptible is the *level of zero amplitude*. The decrease of amplitude with depth is illustrated in a more generalized way in Fig. 4.3. If the ground temperature curve with depth does not go below freezing at any point at its warmest extreme, but does at its coldest extreme, a *seasonal frost* condition exists. If the ground temperature curve remains below freezing over a portion of its length at its warmest extreme, a *permafrost* condition exists.

Below the level of zero annual amplitude, as illustrated in Fig. 4.3, the *geothermal gradient* may be observed, corresponding to the continuous flow of heat occurring from the earth's hot interior toward the relatively cool earth's surface. Within the upper several hundred feet of the earth's mantle, of most concern to human activity, the geothermal gradient is a function of the thermal conductivity and of the unit rate of heat flow:

$$i_g = \frac{q}{k}$$

where

i_g = geothermal gradient,
q = quantity of heat flowing through a unit area per unit time, and
k = thermal conductivity.

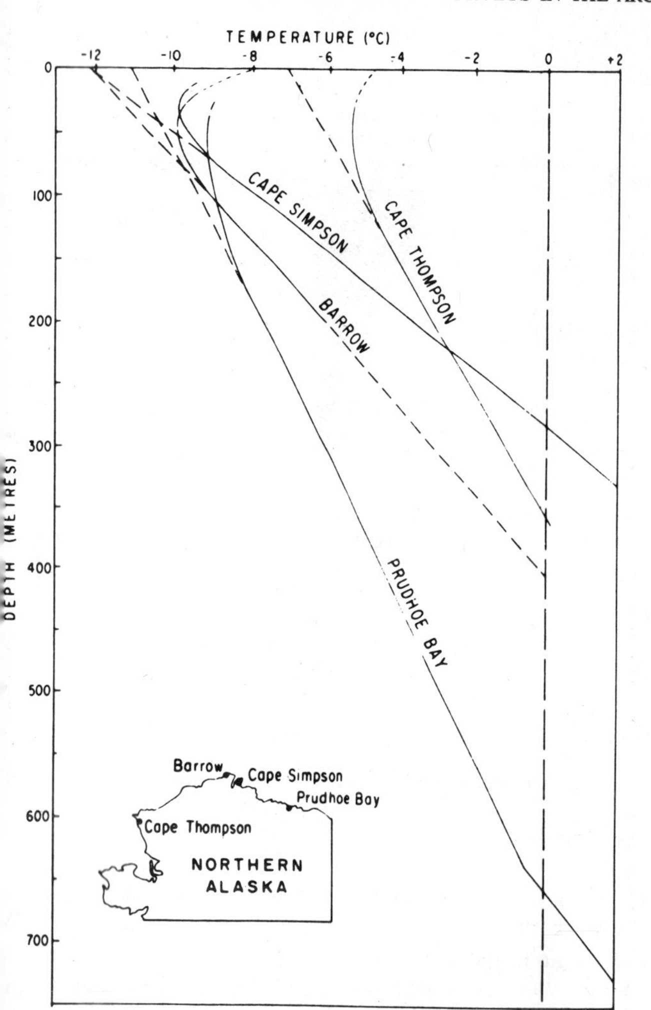

. Generalized profiles of measured temperature on the Alaskan arctic
lid lines). Dashed lines represent extrapolations (Gold and Lachenbruch
eproduced from *Permafrost: Second International Conference*, p. 10,
permission of the National Academy of Sciences, Washington, D.C.).

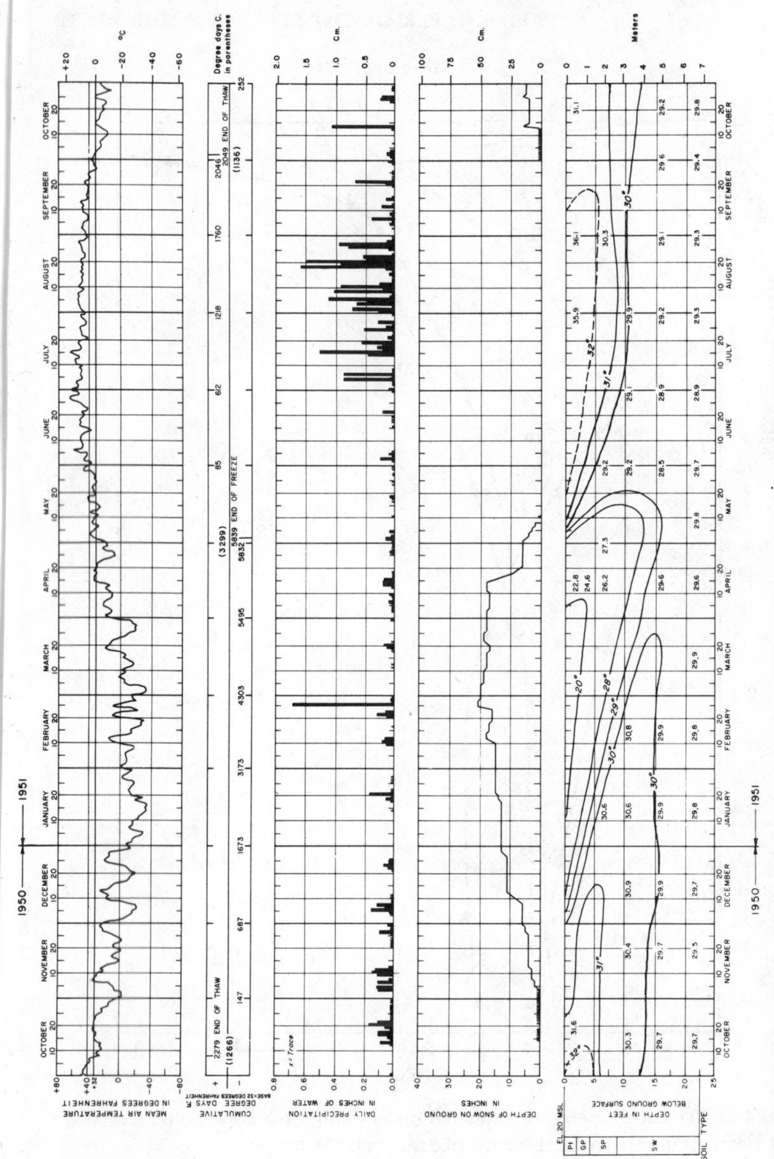

Fig. 4.1. Meteorological data and ground isotherms, Kotzebue, Alaska (Aitken 1965).

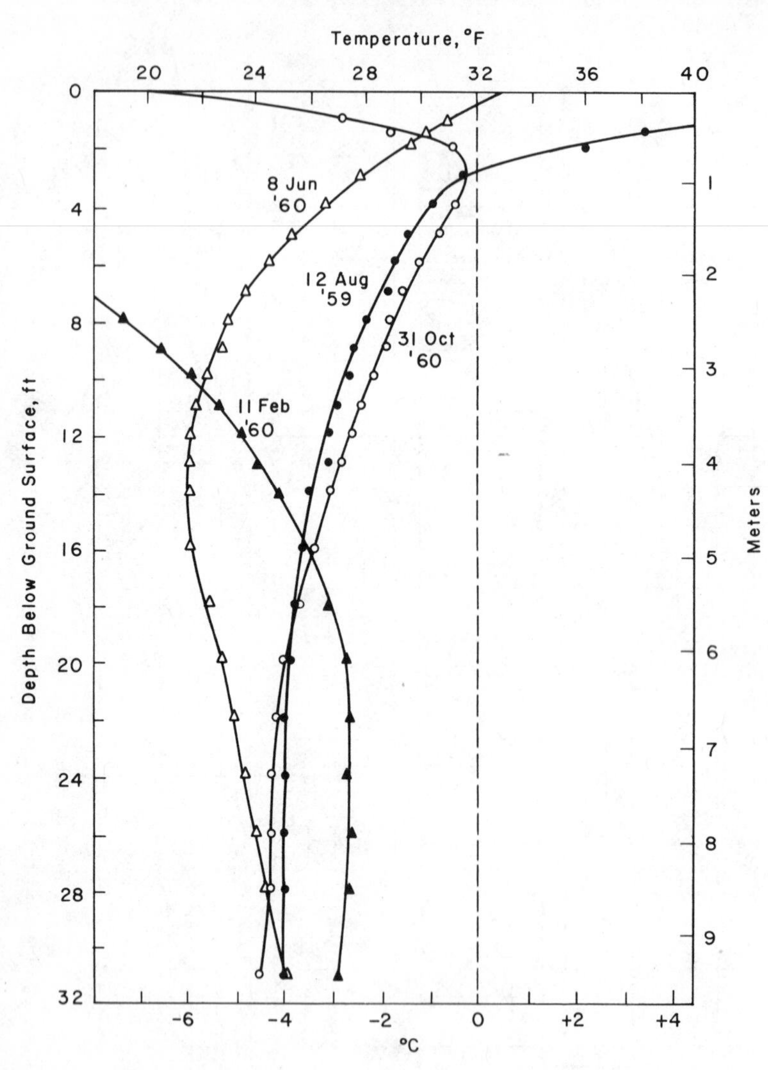

Fig. 4.2. Typical temperature gradients under permafrost conditions, Kotzebue Air Force Station, Alaska (Linell, Lobacz, et al. 1980).

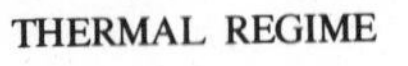

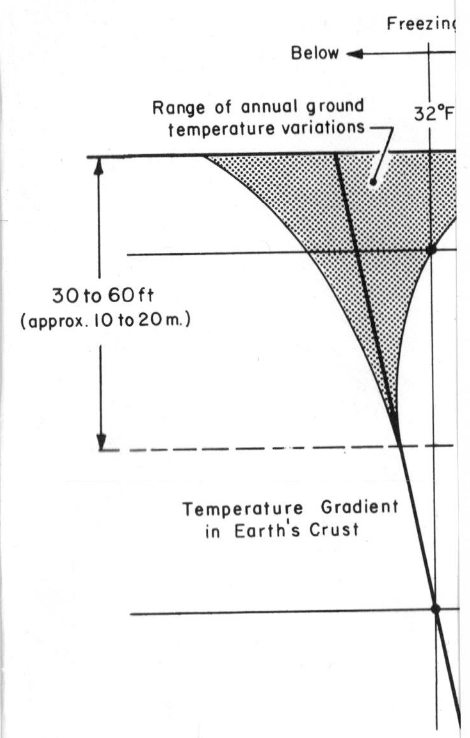

Fig. 4.3. Typical temperature gradients in th
1980).

According to Terzaghi (1952), q in nonp
between 2 x 10^{-6} cal/cm²/sec (7.4 x 10^{-6}
volcanic regions and less than 1 x 10^{-6} ca
sec, 0.042 W/m²) in nonvolcanic areas. T
when heat is generated by chemical actic
deposits. Rates of temperature increase v
measured from 1°C per 12 m (approx. 1°F
100 m (approx. 1°F per 180 ft) of depth.
an average value of about 1°C for every 3
depth (average i_g = 0.033 (°C) m^{-1} = 0.01
reported that geothermal gradients in per
100 ft (9 to 30 m) range from about 1°F p
per 100 ft (1°C per 55 m). Figure 4.4 s

Fig. 4
coast
1973.
with t

typical North American permafrost locations. The geothermal gradient is also affected by long-term changes in the surface temperature of the ground, but the rate of response is slow. As shown in Fig. 4.4, climatic change in northern Alaska since the mid-19th century has produced changes in geothermal gradient down to depths of 75 to 150 m (approx. 250 to 500 ft); Lachenbruch and Marshall (1969) deduced from such data that about a 4°C (7.2°F) rise in the mean annual surface temperature occurred at Barrow, Alaska, after the middle of the 19th century and that a 1°C (1.8°F) cooling had occurred in the decade prior to their report.

Terzaghi (1952) concluded that a sudden rise in mean annual surface temperature of 2°C (3.6°F), from −1°C to +1°C (30.2°F to 33.8°F), would result in the thawing of 15 m (49 ft) of permafrost from the top down in less than 100 years, and only about 2 m (6.6 ft) from the bottom up in the same time, from geothermal heat, if ice occupied 30 percent of the total volume of frozen sediment.

Geothermal heat may be of considerable importance where the resulting soil temperatures are sufficiently high to modify significantly ground freezing conditions and snow and ice accumulations or where heated groundwater or hot springs may occur. The temperature of oil extracted from considerable depths can cause substantial engineering problems in the wells and in surface pipelines and storage facilities where thaw-unstable permafrost exists. For example, at Prudhoe Bay, Alaska, oil is produced from a depth of about 10,000 ft (3000 m) at a temperature of about 195°F (90°C); it passes through up to 2000 ft (600 m) of permafrost before reaching the surface and is transported by pipeline at temperatures in the vicinity of 158 to 176°F (70 to 80°C).

Under average conditions, however, the temperature conditions produced in the upper 60 ft (20 m) of the ground by the direct effect of the surface heat-energy exchange are usually much more directly important than geothermal heat flow. Disregarding geothermal heat flow and heat generated in the ground by chemical or other processes, the thermal regime of the upper part of the earth's surface is the result of the heat flow which takes place into and out of the air–earth interface, assumed to be a plane of infinitesimally small thickness. The principal sources of heat flow to or from this ground-surface layer are as follows:

(i) Net radiation. This is affected by latitude, season, cloud cover, humidity, atmospheric dust, surface type and albedo, and shading and aspect with respect to the position of the sun.

(ii) Heat transfer between air and ground by conduction and convection. This is affected by such factors as wind speed, surface characteristics, and season.

(iii) Evapotranspiration. Both the evaporation of moisture directly from the ground and its transpiration by vegetation involve heat-energy exchange.

The temperature of the ground may also be influenced by the direct heating or cooling effects of falling rain or snow. In addition, heat may be transported vertically and laterally within the ground by moving ground moisture. Finally, the thermal properties of the ground have an effect on surface temperatures.

Changes in any of the influencing factors, such as damage to surface cover by transient passage, shading the surface, changing the snow accumulation patterns by erection of a structure, removing the vegetation, or covering it with soil or pavement, will change the thermal regime and in permafrost areas will result in either raising or lowering the permafrost table. Under unfavorable ground or temperature conditions, permafrost degradation may continue year after year. If degradation occurs in ground which contains excess ice, subsidence of the surface will occur, and progressive erosion and destruction of terrain may develop. These effects are discussed in more detail in Chapter 11.

Rigorous analytical evaluation of the effects of the many variables for predicting the thermal conditions which will result under specific combinations of conditions is extremely complex. No simple and accurate correlation of air temperatures, radiation, and evapotranspiration with surface temperatures for all conditions exists, although Lunardini (1978), Dempsey and Thompson (1969), and others have offered analyses applicable for certain conditions such as pavements. For engineering design, a semi-empirical approach involving only air temperatures is usual.

The most important weather factors influencing the penetration of freezing temperatures into the ground are the air temperatures with respect to 0°C and the duration of the freezing season. The *freezing index* provides a measure of the combined effects of these factors. It is the number of degree days between the highest and lowest points on a curve of cumulative degree days of air temperature versus time taken through one complete freezing season (the season when the average daily temperature generally remains below freezing). It is used as a measure of the combined magnitude and duration of sub freezing temperatures occur-

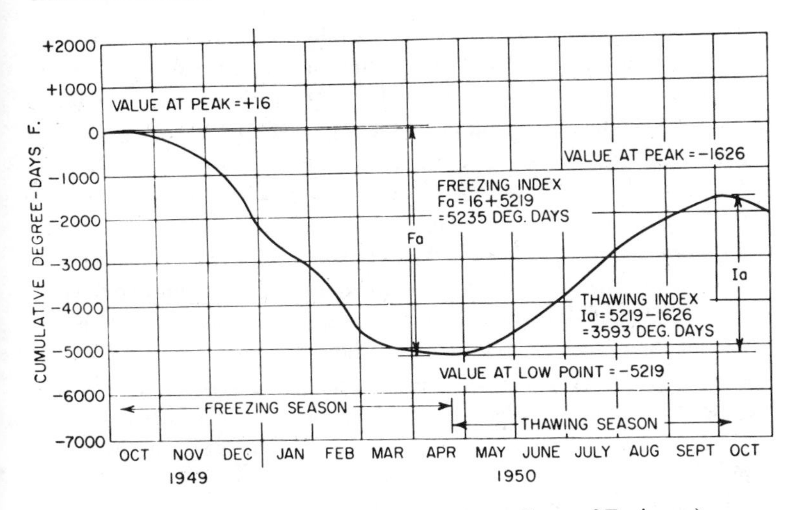

Fig. 4.5. Freezing and thawing indexes (U.S. Army Corps of Engineers).

ring during any given freezing season. The index determined for air temperatures at 4.5 ft (1.4 m) above the ground is commonly designated as the *air freezing index,* and that determined for temperatures immediately below a surface is known as the *surface freezing index*. The freezing index is illustrated in Fig. 4.5.

The *air thawing index,* used as a measure of the combined duration and magnitude of above-freezing temperatures occurring during the complete thawing season, is similarly computed for air temperatures above freezing and is also illustrated in Fig. 4.5.

Ground-surface conditions have a major effect on heat flow into or out of the ground and on the occurrence and thermal stability of permafrost. Observations have established that mean annual air temperatures in permafrost areas are usually at least 1 to 6°C (2 to 11°F) below freezing although exceptions may occur. The 1 to 6°C differential is attributed to the effects of such factors as snow cover, vegetation, time-varying ground thermal properties, relief, slope and orientation of the surface, and surface and subsurface drainage (Gold and Lachenbruch 1973). Surface conditions in the arctic and subarctic may range from forests, muskeg and tundra to bare soil or bedrock, and from water-covered in summer to snow- and ice-covered in winter. It is well known that such materials as moss, lichen, and peat provide excellent thermal barriers. Vegetation also intercepts sunlight before it can reach the ground and dissipates energy to the atmosphere by reflection, reradiation, and evapotranspiration. Vegetation significantly affects the conductive and convective processes of heat exchange between air and ground. In the subarctic, dense tree growth in combination with underlying brush, grass, moss, and other low plant growth, creates conditions favorable for low ground temperatures and possibly permafrost; the lower vegetation may by itself be insufficient to create conditions favorable for permafrost. The possible combinations of vegetation, snow cover, relief, drainage, and other factors are almost infinite, even within a relatively small area of terrain. Nevertheless, the resulting variations in ground temperature alone in undisturbed ground within a given sitting area are not necessarily significant in engineering design except where the variations are substantial, as at the edge of and under a water body, or where they determine the presence or absence of permafrost. The ground temperature changes which may be caused by development or construction can be of major significance, however.

Ratios of ground temperature to air temperature, or *n-factors,* have been empirically developed for various common types of surface conditions. These can be used to convert air freezing and thawing indexes to approximate ground surface indexes for design purposes. Table 4.1 shows *n*-factors currently used by the U.S. Army Corps of Engineers. It should be recognized that because these factors provide no separate

TABLE 4.1.
n-Factors for freeze and thaw (ratio of surface index to air index)

Type of surface[a]	For freezing conditions	For thawing conditions
Snow surface	1.0	—
Portland cement concrete	0.75	1.5
Bituminous pavement	0.7	1.6–2+[b]
Bare soil	0.7	1.4–2+[b]
Shaded surface	0.9	1.0
Turf	0.5	0.8
Tree-covered	0.3[c]	0.4

[a] Surface exposed directly to sun or air without any overlying dust, soil, snow, or ice, except as noted otherwise, and with no building heat involved.

[b] Use lowest value except in extremely high latitudes or at high elevations where a major proportion of summer heating is from solar radiation.

[c] Data from Fairbanks, Alaska, for single season with normal snow cover permitted to accumulate.

Source: Linell, Lobacz, et al. 1980.

consideration of such variables as wind velocity, percentage cloudiness, or heat exchange by radiation, and because the heat contributed by solar radiation in summer can be an appreciable part of the total heat input, considerable imprecision may be involved unless the n-factors used are based on measurements made at or near the site under consideration over a number of years. Even then, it should be understood that the n factor for a given site may vary appreciably from year to year.

The basic thermal properties of earth materials controlling the rates of flow of heat in the ground and depths of seasonal freeze and thaw are the *specific heat,* the *volumetric heat capacity,* the *volumetric latent heat of fusion* of the moisture component, and the *thermal conductivity.*

The *specific heat,* c, is the heat absorbed (or given up) by a unit weight of a substance when its temperature is increased (or decreased) by 1°. The specific heat of most dry soils near 32°F (0°C) may usually be assumed constant at 0.17 Btu/lb F (0.17 calories/gram/°C, 0.71 kJ/kgK) for practical applications.

Volumetric heat capacity, C, is the quantity of heat required to change the temperature of a unit volume by one degree

For dry soils $$C = c\gamma_d$$

For moist, unfrozen soils $$C_u = \gamma_d \left(c + 1.0 \frac{w}{100}\right)$$

For moist, fully frozen soils $$C_f = \gamma_d \left(c + 0.5 \frac{w}{100}\right)$$

Average value for moist soils $$C_{avg} = \gamma_d \left(c + 0.75 \frac{w}{100}\right)$$

where γ_d is the dry-unit weight of soil and w is the moisture content of soil in percentage of dry weight.

The *volumetric latent heat of fusion,* L, is the quantity of heat required to melt the ice (or freeze the water) in a unit volume of soil without a change in temperature

$$L = \gamma_d \frac{w}{100} L_s$$

where L_s is the gravimetric latent heat of fusion of water-ice (144 Btu/lb or 335 x 10^3J/kg).

Thermal conductivity, K, is the quantity of heat flow in a unit time through a unit area of a substance caused by a unit thermal gradient.

The thermal conductivity of soil is dependent upon density; moisture

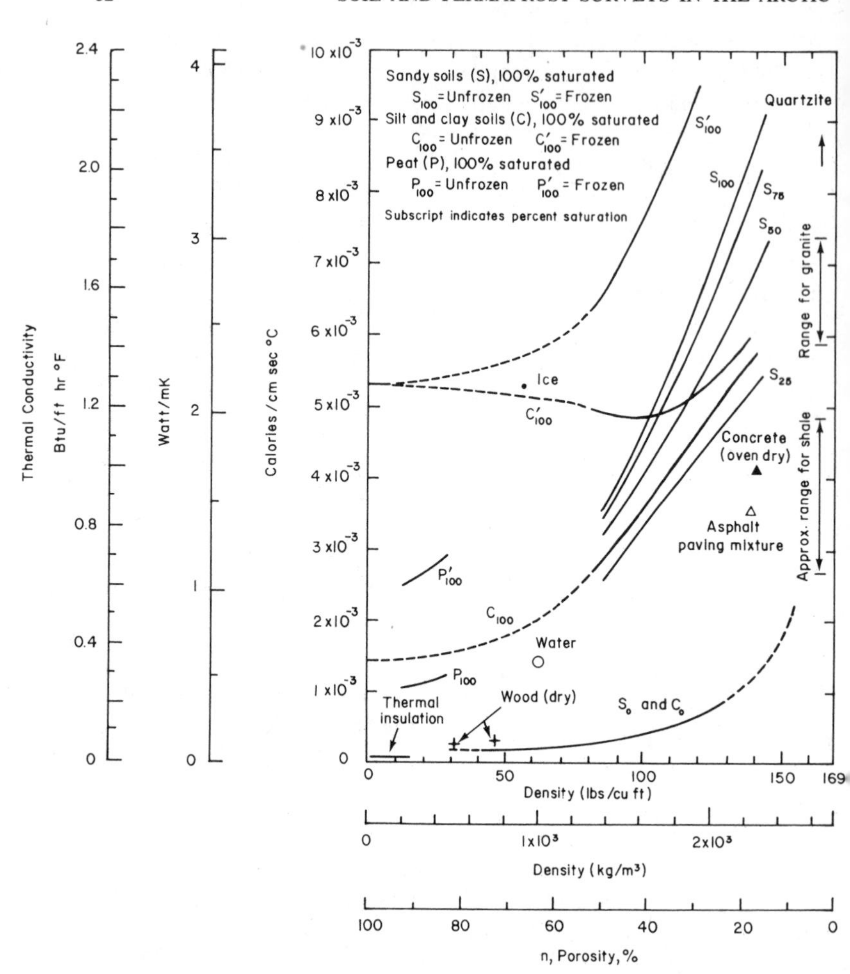

Fig. 4.6. Thermal conductivity versus density and porosity of typical materials. The density values of soils and insulations are based on dry unit weights. (Modified from Terzaghi 1952).

content; temperature; particle shape; solid, liquid, and vapor constituents; and whether the pore water is frozen or unfrozen. Typical values of thermal conductivity versus dry unit weight are shown in Fig. 4.6 for frozen and unfrozen granular soils, silt and clay soils, and peat. Effects of the degree of saturation are also shown. Values for some other common materials are shown for comparison. The values for soils were published by Kersten (1949) and have been widely reprinted by others. Data from Kersten, as presented by Sanger (1966), are shown in Fig. 4.7.

Specific heat and thermal conductivity can be combined to give *thermal diffusivity, a,* which is an index of the facility with which a material will undergo temperature change

$$a = \frac{K}{C} .$$

Typical values of thermal diffusivity for different materials are shown in Table 4.2. Because the diffusivity of ice is much higher than that of water, saturated, frozen soil can change temperature much more readily than equivalent unfrozen soil under equivalent conditions, as long as no change of state is involved.

Because of the large amount of heat released or absorbed at the freezing temperature as latent heat, a discontinuity or pause develops in the curve of temperature versus time for a soil subjected to a cooling or warming condition, as it passes through the transition between unfrozen and frozen or vice versa. This is illustrated in Fig. 4.8, which shows cooling curves for soil specimens subjected to sudden changes of external temperature. After crystallization was initiated during cooling, further cooling was delayed for sufficient time for the heat released by the phase change of the water to be conducted away. Virtually all the free moisture in the silts and the sand froze at 32°F (0°C), but progressive freezing of additional amounts of water occurred with decreasing temperature in the clay, causing absence of a horizontal segment in the cooling curve. Supercooling occurred in each of the soils before initial crystallization.

TABLE 4.2.
Typical thermal diffusivity values

	$cm^2\ sec^{-1}$		
Copper	1133×10^{-3}	Dense saturated sand	$8\pm \times 10^{-3}$
Iron	173×10^{-3}	Soft saturated clay	$4\pm \times 10^{-3}$
Quartzite	45×10^{-3}	Fresh snow	3.3×10^{-3}
Granite	15×10^{-3}	Dry soil	$2.5\pm \times 10^{-3}$
Ice	11.2×10^{-3}	Water	1.4×10^{-3}

Source: Terzaghi 1952.

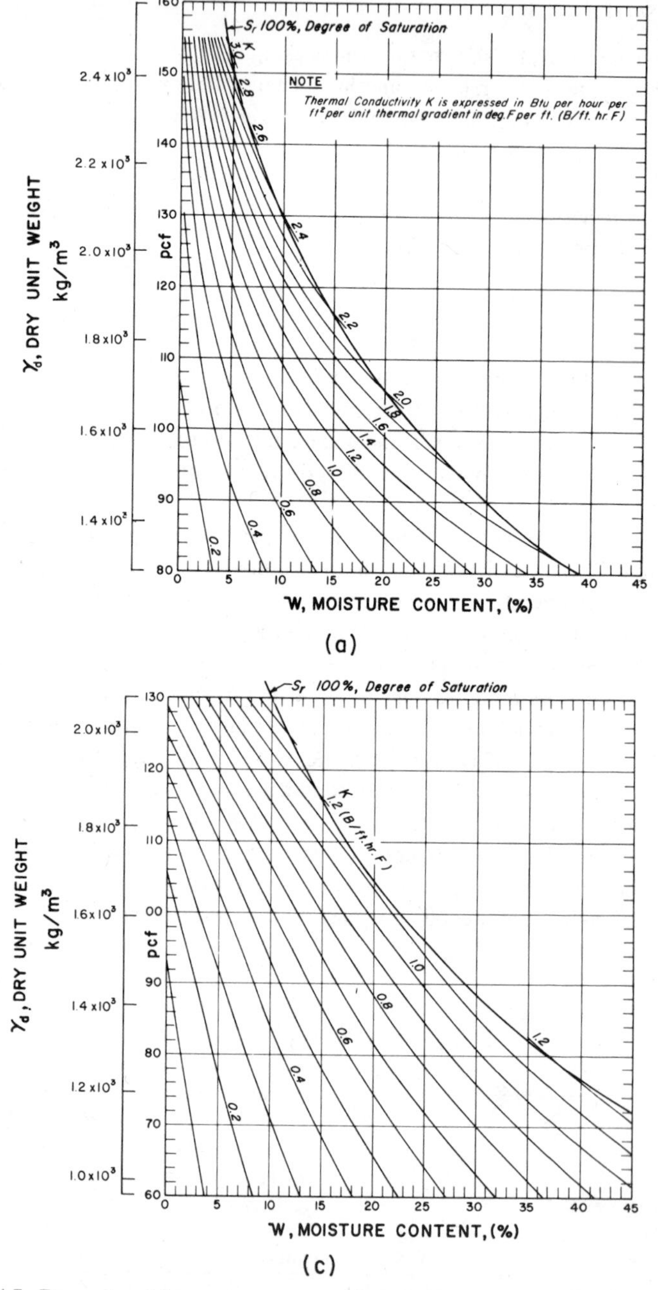

Fig. 4.7. Dry unit weight, water content, and coefficient of thermal conductivity: (a) coarse-grained, frozen soils; (b) coarse-grained, unfrozen soils; (c) fine-grained, frozen soils; and (d) fine-grained, unfrozen soils (Sanger 1966 after

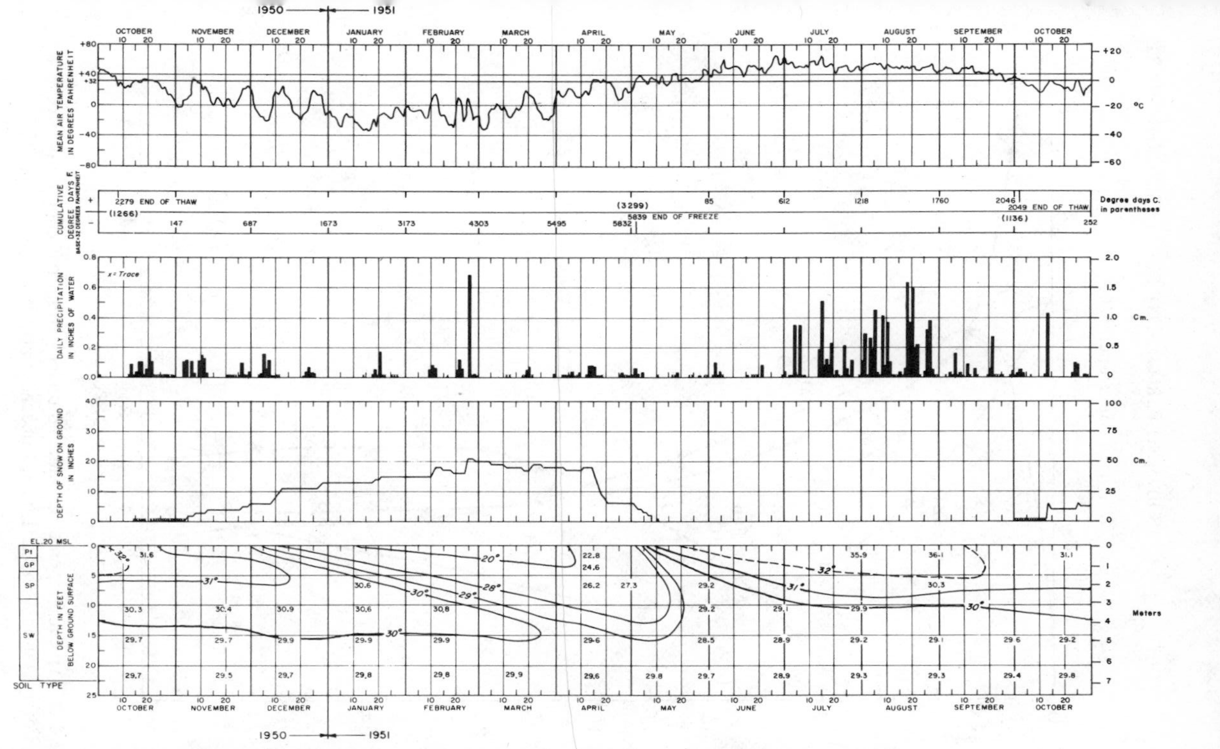

Fig. 4.1. Meteorological data and ground isotherms, Kotzebue, Alaska (Aitken 1965).

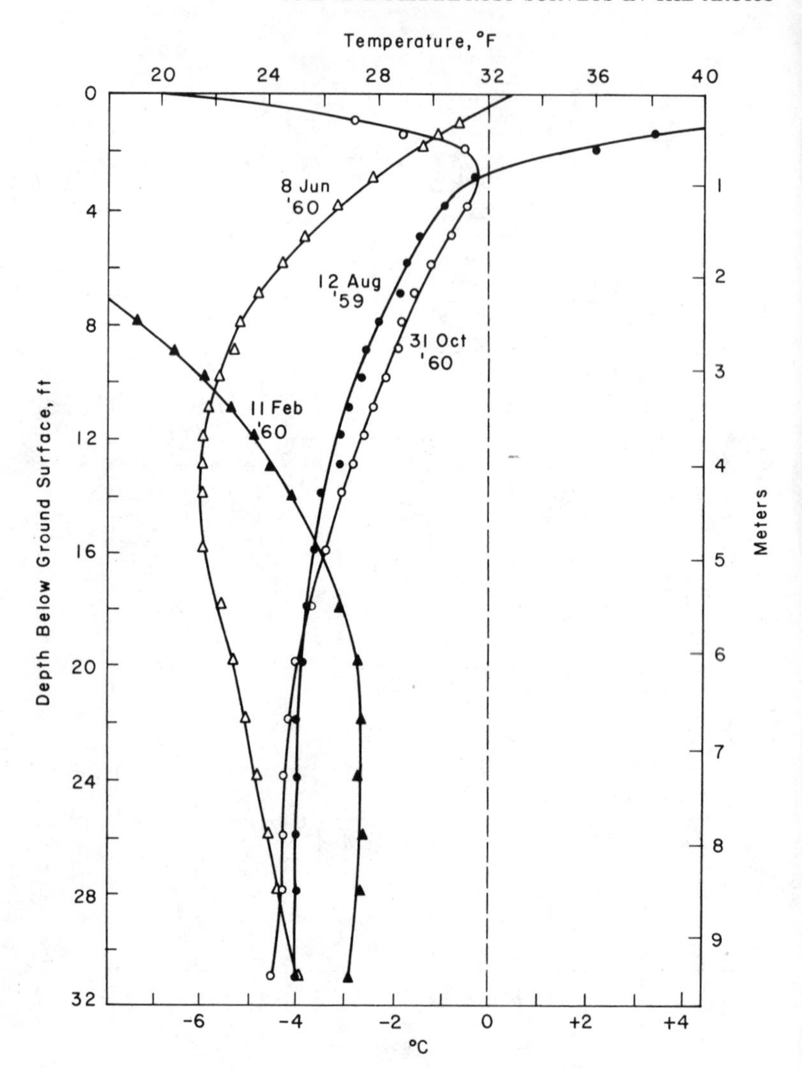

Fig. 4.2. Typical temperature gradients under permafrost conditions, Kotzebue Air Force Station, Alaska (Linell, Lobacz, et al. 1980).

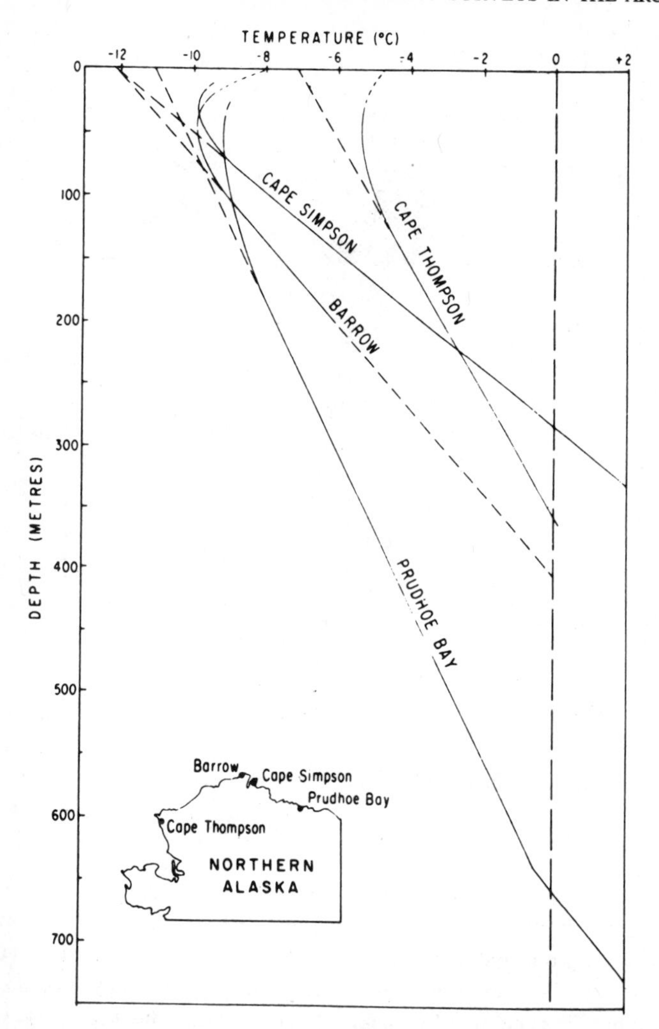

Fig. 4.4. Generalized profiles of measured temperature on the Alaskan arctic coast (solid lines). Dashed lines represent extrapolations (Gold and Lachenbruch 1973. Reproduced from *Permafrost: Second International Conference*, p. 10, with the permission of the National Academy of Sciences, Washington, D.C.).

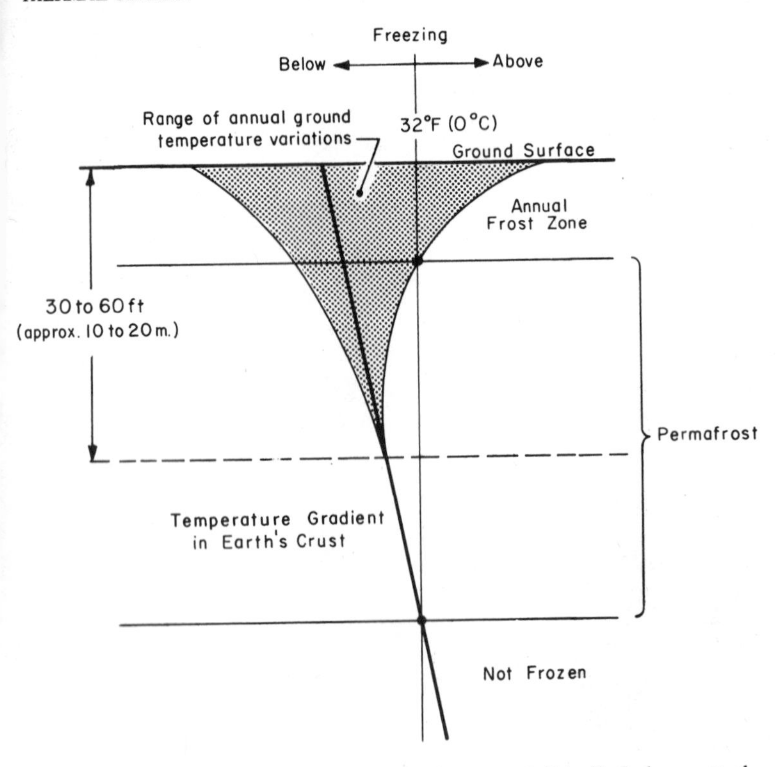

Fig. 4.3. Typical temperature gradients in the ground (Linell, Lobacz, et al. 1980).

According to Terzaghi (1952), q in nonpermafrost materials may range between 2 x 10^{-6} cal/cm²/sec (7.4 x 10^{-6} BTU/ft²/sec, 0.084 W/m²) in volcanic regions and less than 1 x 10^{-6} cal/cm²/sec (3.7 x 10^{-6} BTU/ft²/sec, 0.042 W/m²) in nonvolcanic areas. The gradient may be increased when heat is generated by chemical action within the strata, as in coal deposits. Rates of temperature increase with depth in rocks have been measured from 1°C per 12 m (approx. 1°F per 40 ft) to less than 1°C per 100 m (approx. 1°F per 180 ft) of depth. Terzaghi (1952) has suggested an average value of about 1°C for every 30 m (approx. 1°F per 55 ft) of depth (average i_g = 0.033 (°C) m^{-1} = 0.018 (°F) ft^{-1}). Stearns (1966) has reported that geothermal gradients in permafrost below depths of 30 to 100 ft (9 to 30 m) range from about 1°F per 35 ft (1°C per 19 m) to 1°F per 100 ft (1°C per 55 m). Figure 4.4 shows geothermal gradients at

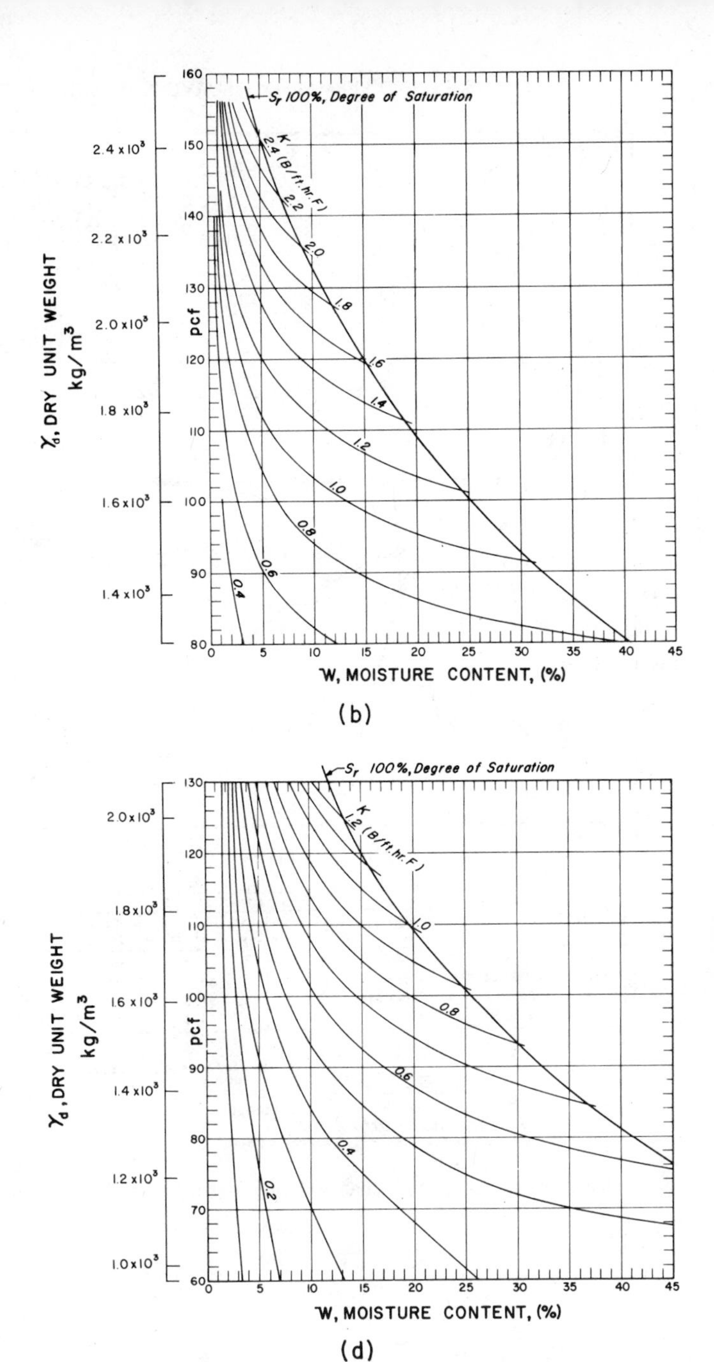

Kersten 1949. Adapted from *Permafrost: Proceedings of an International Conference*, 1966, with the permission of the National Academy of Sciences, Washington, D.C.).

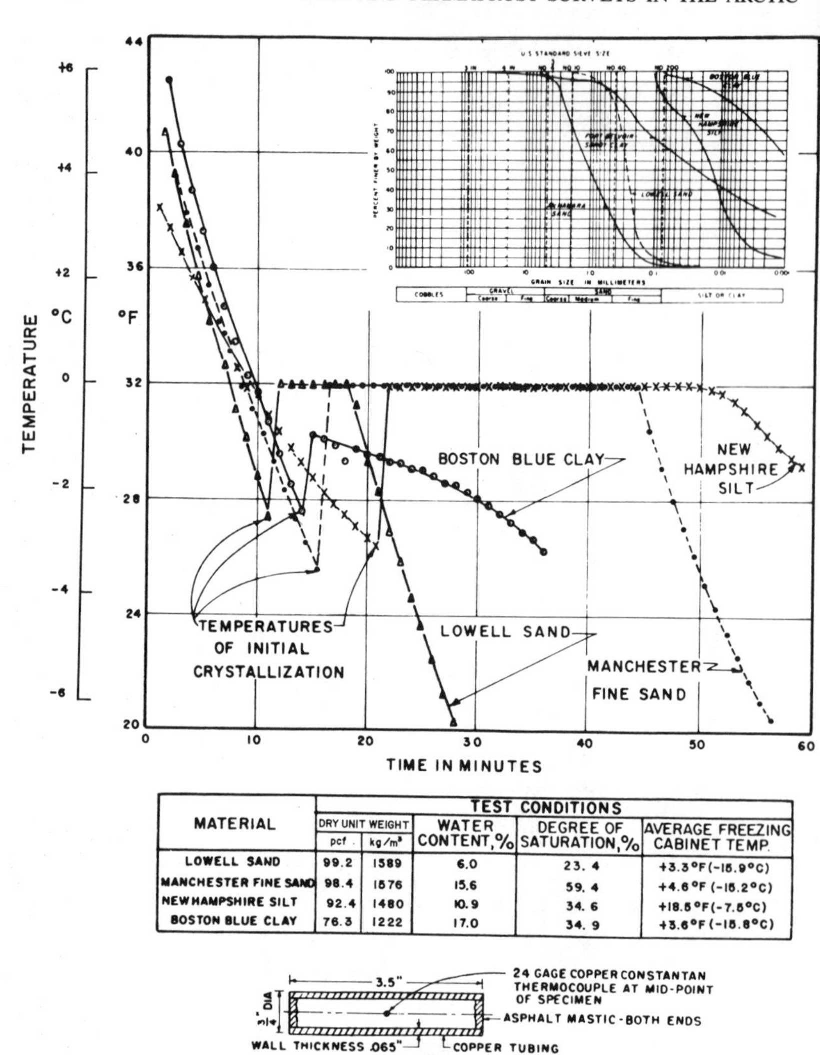

MATERIAL	TEST CONDITIONS				
	DRY UNIT WEIGHT		WATER CONTENT, %	DEGREE OF SATURATION, %	AVERAGE FREEZING CABINET TEMP.
	pcf	kg/m³			
LOWELL SAND	99.2	1589	6.0	23.4	+3.3°F (-15.9°C)
MANCHESTER FINE SAND	98.4	1576	15.6	59.4	+4.6°F (-15.2°C)
NEW HAMPSHIRE SILT	92.4	1480	10.9	34.6	+18.5°F (-7.5°C)
BOSTON BLUE CLAY	76.3	1222	17.0	34.9	+3.6°F (-15.8°C)

Fig. 4.8. Typical plots of temperature change during freezing (Linell and Kaplar 1959).

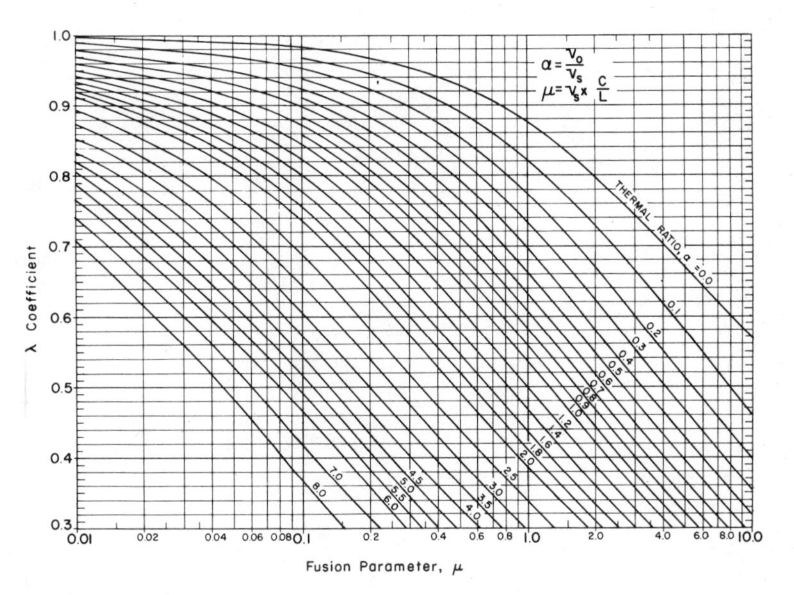

Fig. 4.9. Lambda coefficient in the modified Berggren formula (U.S. Army/U.S. Air Force 1966c, modified from Aldrich and Paynter 1953).

In planning and executing surveys it may often be necessary to anticipate the depths of seasonal freeze or thaw which will be encountered. The depth of freeze or thaw in soil depends upon the ground-surface temperature, the thermal properties of the soil mass, and the average initial temperature of the soil at the start of the freezing or thawing season. For purposes of calculation, the average initial ground temperature is generally assumed to equal the mean annual temperature of the site for which depths of freeze and thaw are being predicted. The ground-surface temperature is represented by the surface freezing or thawing index. The complications introduced by the latent heat of fusion are most commonly handled by using the modified Berggren equation (Aldrich and Paynter 1953) to estimate depths of seasonal freezing in seasonal frost areas and depths of seasonal thaw in permafrost areas:

$$X = \gamma\sqrt{48KnF/L} \quad or \quad X = \gamma\sqrt{48KnI/L}$$

where, using either all Customary units or all SI or derived SI units,

X = depth of freeze or thaw, ft or m,

K = thermal conductivity of the soil, Btu/ft hr °F, or

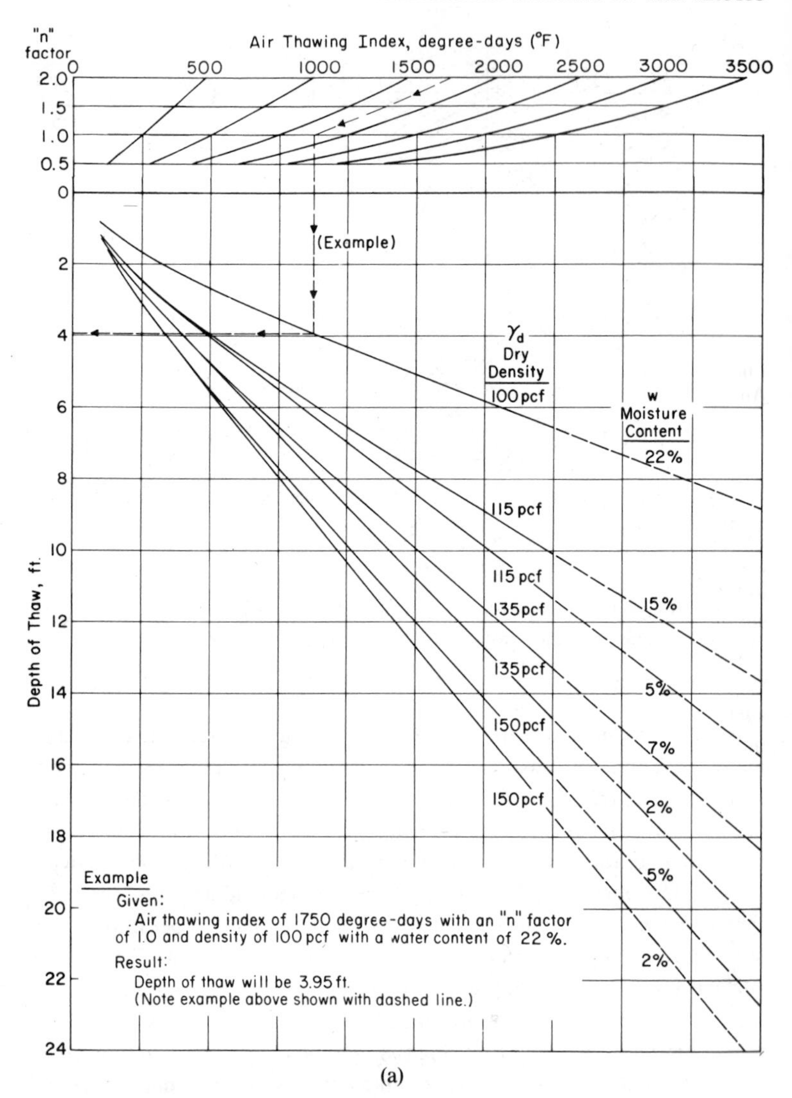

(a)

Fig. 4.10. Approximate depth of thaw or freeze versus air thawing (a) or air freezing index (b) and n factor for various homogeneous soils (Linell, Lobacz, et al. 1980).

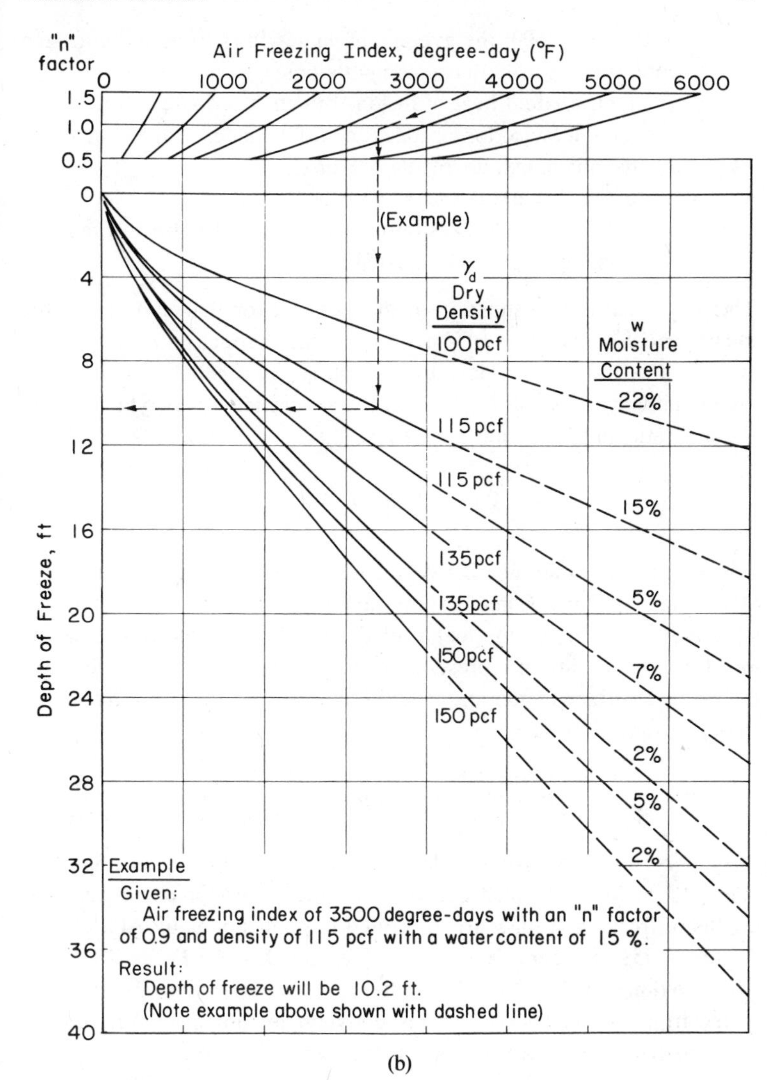
"n" factor
Air Freezing Index, degree-day (°F)
0
1000
2000
3000
4000
5000
6000
1.5
1.0
0.5
0
4
8
12
16
20
24
28
32
36
40
Depth of Freeze, ft
(Example)
γ_d
Dry Density
100 pcf
115 pcf
115 pcf
135 pcf
135 pcf
150 pcf
150 pcf
w
Moisture Content
22%
15%
5%
7%
2%
5%
2%
Example
Given:
Air freezing index of 3500 degree-days with an "n" factor of 0.9 and density of 115 pcf with a water content of 15 %.
Result:
Depth of freeze will be 10.2 ft.
(Note example above shown with dashed line)

(b)

watt/m°K × 3600; the average of values for frozen and unfrozen conditions is used in the computations,

L = volumetric latent heat of fusion, Btu/cu ft or joule/cu m,
n = conversion factor for air index to surface index, dimensionless,
F = air freezing index, degree-days F or C,
I = air thawing index, degree-days F or C, and
γ = a coefficient which takes into consideration the effect of temperature changes in the soil mass.

The γ coefficient is a function of the freezing (or thawing) index, the mean annual temperature of the site, and the thermal properties of the soil. Freeze-thaw of low-moisture-content soils in the lower latitudes is greatly influenced by this coefficient. It is determined by two factors, the thermal ratio and the fusion parameter.

$$\text{Thermal ratio } \alpha = \frac{v_0}{v_s}$$

where v_0 is the difference between the mean annual site temperature (MAT) and 32°F or 0°C and v_s is the surface freezing (or thawing) index, nF (or nI), divided by the length of the freezing (or thawing) season, nF/t (or nI/t). This term represents the temperature differential between the average surface temperature and 32°F or 0°C for the entire freeze (or thaw) season.

$$\text{Fusion parameter } \mu = v_s \times \frac{C_{avg}}{L}$$

where terms are as previously defined. Figure 4.9 shows γ as a function of α and μ. More detailed guidance in procedures for calculating the depths of freeze and thaw are presented in publications by Aldrich and Paynter (1953), Sanger (1966), and U.S. Army/U.S. Air Force (1966_c).

Approximate values of thaw or freeze penetration can also be estimated from Fig. 4.10(a) or (b), respectively, for homogeneous soils of the density and moisture content ranges there represented.

Thaw penetration observed during an arctic summer is illustrated in Fig. 4.11. In mid-June, when daily air temperatures rose above freezing, thaw depth already exceeded 1 ft (0.3 m) because of ground heating by direct solar radiation not reflected in the air temperature measurements. Thaw penetration was very rapid in the last part of June with the sun at its maximum elevation and air temperatures near their warmest. In July and early August thaw penetration slowed as the strength of the solar

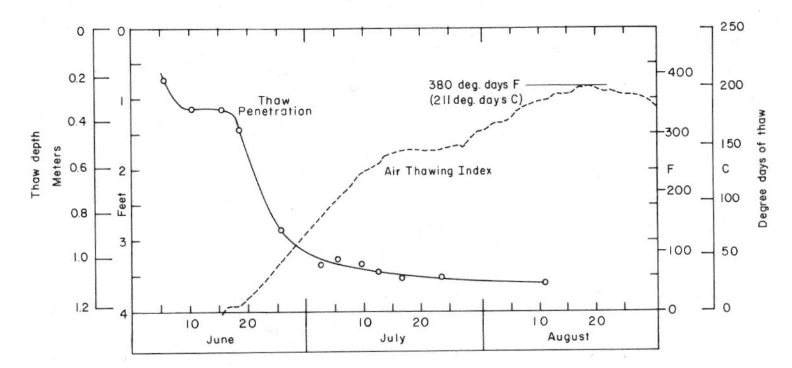

Fig. 4.11. Thaw progression under undisturbed surface, Camp Tuto, Greenland. Soil data: average classification SC, clayey sand; dry unit weight 120 to 124 lb/ft^3 (1920–1990 kg/m^3); moisture content 8 to 12 percent; essentially no vegetative cover on surface (ACFEL 1963).

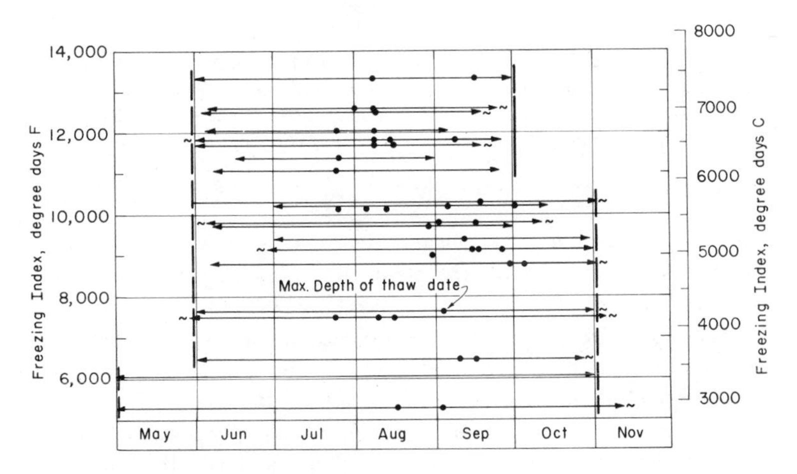

Fig. 4.12. Period of thaw versus freezing index. Based on soundings at 38 locations in northern Canada (Sebastyan 1966. Adapted from *Permafrost: Proceedings of an International Conference,* p. 388, with the permission of the National Academy of Sciences, Washington, D.C.).

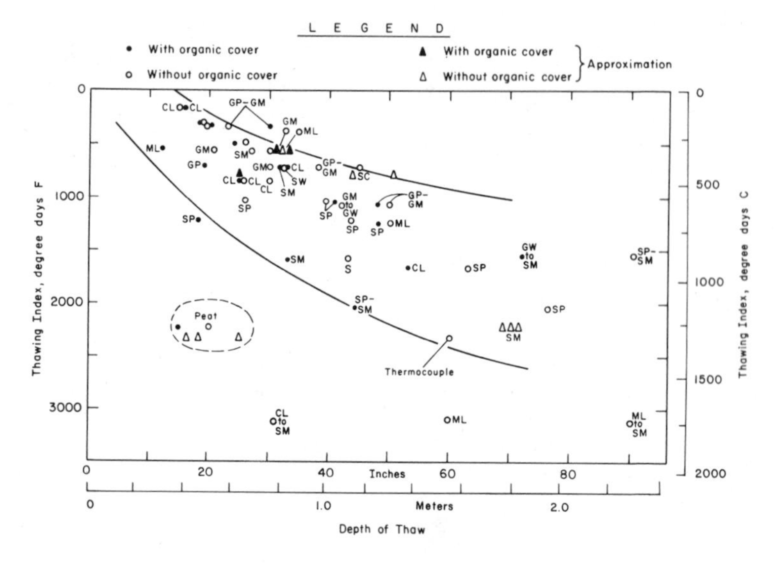

Fig. 4.13. Depth of thaw versus air thawing index for unpaved surfaces. Based on soundings at 62 locations in the Canadian North (Sebastyan 1966. Adapted from *Permafrost: Proceedings of an International Conference,* p. 388, with the permission of the National Academy of Sciences, Washington, D.C.).

input decreased rapidly, air temperatures cooled somewhat more gradually from their peak during the period from June 20 to July 15, and resistance to thaw advancement into the ground increased. Although not illustrated, refreezing of the thawed layer had already started both from the top down and from the bottom up by 1 September.

Figure 4.12 shows the length of the period of thaw at thirty-eight locations in northern Canada, plotted against freezing index as a measure of the site temperature conditions. Dates of maximum depth of thaw are seen to be in August and September at the colder locations, extending into October and even November at warmer, more southerly locations. Figure 4.13 shows observed depths of seasonal thaw under unpaved surfaces in permafrost areas of northern Canada. Thaw depths one and one-half to two times greater may occur under bituminous pavements.

5. Patterned ground

One of the more striking features of arctic terrain is that of patterned ground. Troll (1958) summarized the state of knowledge of patterned ground forms up to the Second World War. Since that time Cailleux and Taylor (1954), Hamelin and Cook (1967), Jahn (1971), and Washburn (1973) have provided additional accounts of the multiplicity of patterned ground forms. Patterned ground is present throughout much of the arctic region and parts of the subarctic as well. To the south, relic forms exist in a wide, indefinite belt across central Siberia, northern Europe, England, southern Canada, the northern United States, and other locations. Patterned ground is also present in the alpine regions of the Northern Hemisphere and even in the high mountains of the equatorial belt.

This volume is concerned primarily with soil and permafrost surveys, and we must point out accordingly that arctic soils are subject to intensive frost action which does not always lead to the development of a geometric surface pattern. There is a mixing effect from frost action within the soils which redistributes the soil components and causes a complex morphology. Mixing is more common where there is a large ice/water content involved in the system, but the process also takes place on well-drained positions where only a small amount of ground ice is present. The problem of erratic distribution of soil properties has been recognized since the early days of arctic soil investigations. At the present time there are at least two approaches to solving the problem: (i) recording the detailed properties at a specific site, with the understanding that the same set of properties will not necessarily be repetitive beyond the sampling site; or (ii) describing the range of properties normally encountered within the mosaic of soil conditions (Douglas and Tedrow 1960). Figure 5.1 shows an example of a tundra soil with a complex morphology resulting from frost action.

On sloping arctic land soil material tends to move downslope causing a highly disturbed set of conditions. The term *solifluction* was originally introduced to describe the slow flowing of masses of soil saturated with water from higher to lower ground (Fig. 5.2). Solifluction associated with frozen ground is sometimes referred to as *gelifluction*. These terms connote the overall mechanism but not for the multiplicity of soil conditions associated with sloping land. Soils on the sloping lands may be con-

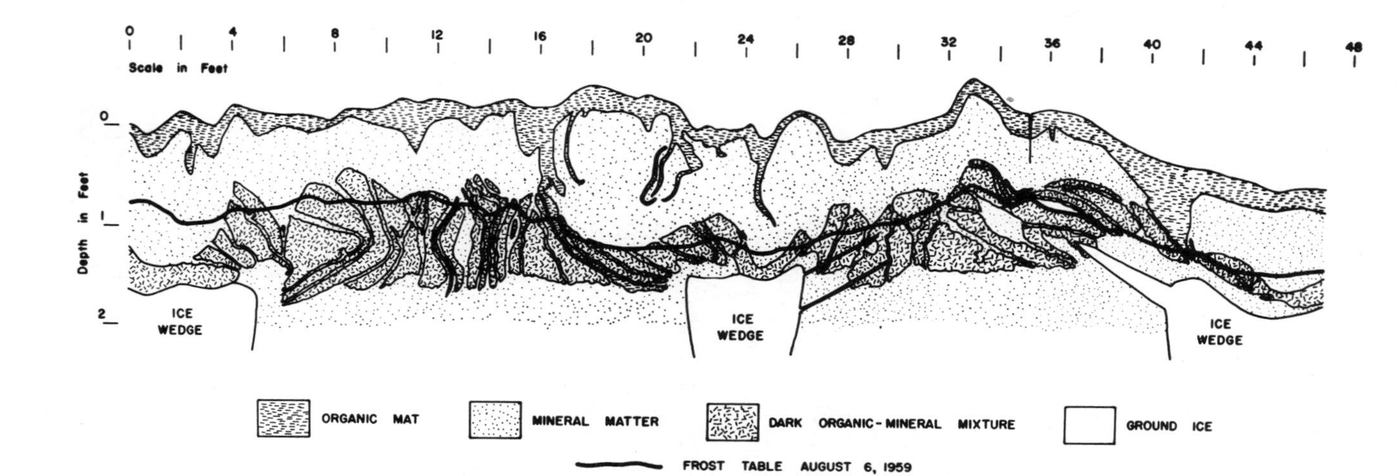

Fig. 5.1. Cross-section of Tundra soil near Barrow, Alaska (Douglas and Tedrow 1960).

Fig. 5.2. Solifluction on Bathurst Island, Northwest Territories, Canada.

sidered somewhat 'Regosolic' in character, in that they display few genetic features, but there is a considerable range of conditions associated with solifluction slopes including that of substrate and drainage. Washburn (1967) determined the rates of near-surface movements in northeastern Greenland on a 10° to 14° gradient and found them to range from 0.6 cm per year on relatively dry spots to 6.0 cm per year on wet spots.

In addition to the processes related to downslope movement of soils there are frost-related features, such as block fields, felsenmere, rock glaciers, talus deposits, and related structures which need to be considered in field classification of soils.

Classifications of patterned ground

Washburn (1956), in trying to bring a better understanding to the complexities of patterned ground configurations, proposed a classification scheme which placed all patterned ground forms into five categories: polygons, circles, nets, steps, and stripes. Each category was then divided into forms having sorted and unsorted features.

Polygons

Perhaps the most frequently described pattern ground form is the ice-wedge polygon, which is one of Washburn's nonsorted polygons. Ice-wedge polygons are widespread, particularly in the wet, level-to-undulating areas of the arctic having some form of gley or Bog soil present. But such polygons are also present in association with other soil conditions and even bedrock. The formation of nonsorted ice-wedge polygons is one of the better understood processes (Lachenbruch 1962). Initially shrinkage cracks develop in the frozen material from seasonal changes in temperature. The newly formed cracks tend to fill with water during summer months and to refreeze during the winter. During subsequent years cracks continue to form, causing a progressive lateral compression. The ice wedges with their 'annual growth' may create such lateral compression forces (Fig. 5.3) that the central area of the polygon is forced upward to form a high-center polygon. In low-center polygons the margins are raised relative to the central areas. Some high-center polygons are believed to eventually convert to low-center polygons (Fig. 5.4).

During the course of field investigations, one should consider the degree of microrelief induced by the polygon formation, whether the polygons are high-center or low-center. The development of microrelief features will alter the moisture regime of the soil, which, in turn, will be reflected by plant communities. Drew and Tedrow (1962) formulated a classification system of ice-wedge polygons which recognized the relative microrelief forms and then correlated the various polygonal forms with the soil varieties.

Sorted polygons form under conditions of intense frost action in which there is a differential displacement of soil particles according to size. One of the most common forms is composed of bare mineral soil with a border of rock fragments (Fig. 5.5). Although sorted polygons usually signify mineral material with minimal soil development, great variations

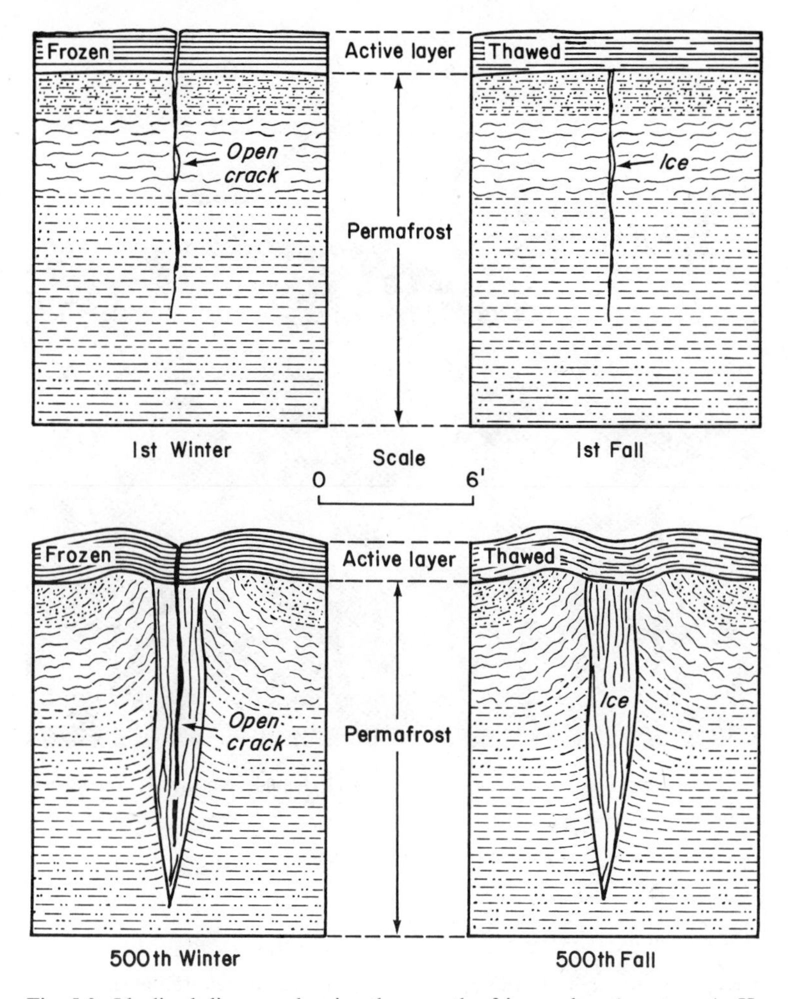

Fig. 5.3. Idealized diagram showing the growth of ice wedges (courtesy A. H. Lachenbruch).

exist. The mineral matrix may consist of material ranging from primarily boulder-size rock to fine clay. Sites may range from the xeric, stony areas of the mountains to the wet mineral soils and bogs of the depressions. Field investigations will show that there are a number of soil conditions associated with sorted polygons.

Fig. 5.4. Low-center polygons, Prince Patrick Island, Northwest Territories, Canada.

Circles and nets

Ground patterns are in the case of circles dominantly circular and in the case of nets neither circular nor polygonal. Circles and nets usually occur as sorted patterns where the matrix contains a prominent rock or gravel component. Through repeated frost action, the coarse fragments tend to become concentrated on the periphery of the structure. High mountains of the arctic have extensive sectors of sorted circles and nets, but such forms also occur at many other locations, such as boulder-strewn lowlands including coarse-textured fluvial deposits.

Nonsorted circles and nets are patterned ground forms which display some geometric form from shrinkage or frost processes but lack a border

Fig. 5.5. Stone rings on Prince Patrick Island, Northwest Territories, Canada.

of stones. Such patterns are usually found on fine-textured materials where the land is nearly horizontal.

Steps

Steps, as defined by Washburn (1973), are steplike patterned ground forms with a downslope border of vegetation or stones embanking an area of bare ground upslope. They are terrace like forms derived from a combination of at least two processes, (i) the primary frost action producing a geometric pattern, such as sorted circles, with (ii) partial downslope movement of the matrix resulting in a steplike condition. With the steplike appearance, the tread of the step is commonly composed of bare soil and the riser consists of turfy material. There are, however, a great many varieties of steps ranging from bare mineral soil structures to hummocky, vegetated terrain. Steps are usually present on sloping sites where the gradient may range from 2 to 15 or more percent.

Stripes

Both sorted and nonsorted patterns occur on slopes ranging from only a few percent to very steep conditions. Sorted stripes exist primarily as alternate stripes of rocks-sod, rocks-fines, or other forms, oriented with the slope of the land. In nonsorted stripes the striped ground pattern results from alternating lines of vegetation and bare ground. The beginning of the formation of the stripes remains somewhat uncertain, but development of the parallel configuration itself is gravity-related. A wide range of soil conditions is associated with the various soil stripes. Some conditions consist of virtually nothing but mineral soil material without genetic soil formation, whereas under some situations the sod-covered soil between the rock stripes shows affinities with Arctic brown soil.

Miscellaneous structures

A number of additional frost-related features should be considered in surveying permafrost and soil conditions. Some of the more important features and their characteristics follow:

Pingos are conical-shaped hills, some up to more than 60 m high. Morphologically, they are ice-cored structures generally having an organic-rich surface.

Palsas are dome-shaped structures up to about 7 m high, usually ice-cored and with an organic surface.

Mounds (frost mounds and peat mounds) are defined for the purpose of this volume as small dome-shaped structures, usually with a core of ice, or ice-lensed soil.

Solifluction lobes and *solifluction terraces* or *benches* are made up of soil material which has moved downslope in lobate or terracelike form. Such forms signify a degree of soil instability.

String bogs (Aapomoor, oapamoor, strangmoor) are present on bog-like terrain and consist of sinuous organic ridges up to 1m or more in height.

TABLE 5.1.

Major types of patterned ground frequently associated with various arctic soils (double xs indicate a greater frequency of the pattern)

Pattern	Rocky, xeric soils—generally shallow	Deep, well-drained soils	Mineral gley soils	Bog soils	Disturbed soils of the steep slopes
Polygons (sorted)	xx	x			
Polygons (unsorted)			xx	xx	
Circles and nets (sorted)	xx	x	x		
Circles and nets (unsorted)			xx		
Steps	x	x	x		x
Stripes (stone)	xx				x
Stripes (sod)	x				x
Mounds (frost)		x	x	x	
Mounds (peat)				x	
String bogs			x	x	

Patterned ground as an indicator of soil conditions

Some investigators have tended to use the patterned ground forms to describe the genetic soil, using such terms as 'rock polygon soils,' 'structure soils' (*Strukturboden*), 'stone ring soils' (*Steinringboden*), and many others. Such an approach should generally be discouraged because a multiplicity of soil conditions may be associated with the various patterned ground forms. Rather than using the patterned ground forms to describe the soil, it is more appropriate to describe the soil in generic terms and to add the patterned ground form to the soil name. Patterned ground, however, is not universally present in the arctic landscape.

Some patterned ground forms may indicate a specific or general soil condition (Table 5.1). For example, sorted or unsorted stone nets are reliable indicators of frost-shattered, lithosolic conditions, or a peat mound will obviously indicate organic terrain. Generally, however, a wide variety of soil conditions may be present with most patterned ground forms. Fedoroff (1966) integrated the generic soil type with the patterned ground form into a soil unit, a realistic procedure.

In addition to the problem of associating patterned ground forms with generic soils in the field. the survey should consider the *degree* of development of the pattern. If, for example, the unsorted ice-wedge polygon has its center elevated only a few centimeters above the surrounding channel, soils and plant spectra will generally be quite different from those polygons with the centers elevated 50 cm or so above the channel.

A number of patterned ground forms may be associated with rather specific terrain conditions—particularly slope and substrate. For example, a string bog nearly always signifies wet, flat terrain, and a field of sorted rock polygons usually indicates well-drained to xeric conditions.

Some forms of patterned ground tend to be associated with certain landscape features and with certain kinds of soil. Accordingly, Table 5.1 has been compiled to show common positions of patterned ground in relation to broad landscape elements.

6. Arctic soil formation and pedologic classification

Soil formation as viewed by the pedologist commonly considers soil to a depth of only 1 to 2 m. Whereas there are some cases in which soil investigations are carried out to much greater depths, especially in the vicinity of cut banks, terrace faces, and landslide areas, such conditions are not widespread. Soil engineers on the other hand by necessity must generally consider soil conditions to greater depths. Because of the tendency to have high order variability of arctic soils with depth, it is unwise to extrapolate pedologic information beyond the actual depth of observations.

Arctic soils have a number of factors and conditions associated with their formation not generally found in other areas of the globe. Chronologically, the majority, though not all, of arctic soils are young in that nearly all glaciated landscapes have been free of glacial ice only within the last 15,000 years or so. In some local sectors the landscape has been ice-free only within comparatively recent time. In addition to disappearance of the ice cover itself, major areas of the arctic have undergone isostatic adjustment resulting in emerging landforms.

A large area surrounding Hudson Bay and extending northward to the arctic islands was depressed well below sea level during Pleistocene time, as was the case in northern Scandinavia and the lower courses of major rivers of the Soviet arctic. Some emerged landforms have accumulated organic-rich sediments, which, coupled with flat terrain and poorly developed drainage patterns, have resulted in sluggish surface drainage and extensive swamplike or muskeg-like conditions. Figure 6.1, from Bathhurst Island, typifies such a condition in which the land has emerged only within about the past 4500 years (Blake 1974). This wet meadow is flat, peaty, and waterlogged, with permafrost at approximately the 25 cm depth. Not all emerged landforms, however, are poorly drained.

Low temperatures of the north create conditions that make some of the soil-forming processes unique. Permafrost is nearly always present, which, coupled with the low ambient and soil temperature regimes, greatly restricts biological processes as well as inorganic reactions within the soil. Permafrost, by virtue of its low temperature, not only affects the rate of chemical reactions within the soil, but also generally produces

Fig. 6.1. View of Polar Bear Pass, Bathurst Island, N.W.T., showing poorly drained, recently emerged landform. Virtually all the area shown is poorly drained Meadow tundra soil and Bog.

impermeable conditions with respect to the potential leaching and migration of substances through frozen layers. During the formation of permafrost, as well as subsequent changes induced through frost action, arctic soils tend to change morphologically. Frost processes frequently cause great irregularity in the soil features—even within short distances. Extrapolation of detailed and minute soil descriptions beyond the point of sampling, therefore, has limitations.

Traditionally investigators have inferred that the arctic soil cover is monolithic in character. Well over a century ago, however, Middendorf (1864) had already recognized that the arctic had high tundra with dry,

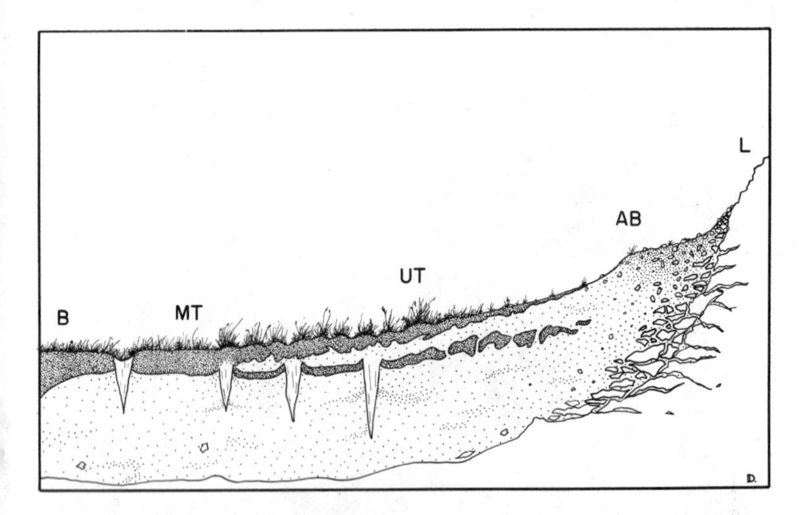

Fig. 6.2. Idealized cross-section of soils of the tundra (low arctic) zone: B, Bog soil; MT, Meadow tundra soil; UT, Upland tundra soil; AB, Arctic brown soil; and L, Lithosol. The cuspate-shaped markings indicate ice wedges.

mineral soils and low tundra with wet, peaty soils. It is now recognized that, in addition to the features described by Middendorf, the arctic soil cover shows great local variation. Also there are latitudinal changes as well as altitudinal changes of sorts within major arctic mountain systems.

Figure 6.2 shows an idealized landscape of the main arctic belt with the range of soil conditions commonly present. The flat, rolling, and, to an extent, hilly sectors have gleyed soils underlain by permafrost. Even though the quantity of precipitation in arctic areas is quite small, there is some downward movement of soil moisture within the active layer. Drainage waters, however, upon reaching the frozen zone within the soil, do not penetrate the permafrost, especially where considerable ground ice is present. As a result of the underlying frozen soil, a gley condition forms within the active layer, reflecting the hydric conditions. Figure 6.3 shows a Tundra (gleyed) soil from northern Alaska. The soil has an irregular organic horizon underlain by an olive brown silt loam. At a depth of 38 to 46 cm, the permanently frozen material is organically enriched. The lower portion of the profile consists of frozen silts and ground ice.

Most Tundra soils are silty in character. Many earlier investigators believed that weathering of rocks to approximately silt size (5 to 50 μm)

Fig. 6.3. Tundra soil near Umiat Lakes, Alaska, showing an organic mat underlain by silt which, in turn, is underlain by buried organic matter (which yielded a ^{14}C date of about 9100 years B.P.). The organic matter rests on ground ice. Scale in feet.

was an arctic phenomenon, but this projection must be considered largely incorrect because it is now established that soil textures range from clays to gravelly sands. From chemical parameters, Tundra soils show a degree of leaching, soil reaction generally being within the moderately to strongly acid range. There are situations, however, particularly with carbonate-bearing rock, in which the soil will be alkaline.

Percentage of base saturation is consonant with the acidity level in the soil.

In depressions and other low-lying areas, Tundra soil will be in a comparatively wetter environment and, accordingly, the organic matter accumulation on the surface will be greater, and the mineral coloration of the soil will tend to be dull gray. Soviet investigators commonly refer to the gray shade as 'dove-colored gray.'' This soil is designated as a *Meadow tundra,* in contrast to the *Upland tundra* soil described in the preceding paragraph.

Differentiation between Upland and Meadow tundra soils is based primarily on the different degrees of wetness within an idealized drainage catena (Tedrow et al. 1958). Targulian (1971) considered Tundra soils according to their degree of development and introduced the term *Homogeneous gleyic soils* to designate those that have developed a sequence of genetic mineral horizons through soil-forming processes. Targulian's approach gives considerable attention to the stage of the developmental cycle of the soil. Thus, for example, he differentiates Homogeneous gleyic soils from *Differentiated gleyic soils,* but it is not always possible to establish whether such differences are time-dependent phenomena or whether they result from climate, parent material, or other differences. Nevertheless, Targulian's thesis is fundamental and attacks the core of the problem, much the same as B. N. Gorodkov did during the 1930s.

Wide expanses of Bog soils are present throughout much of the arctic —primarily where terrain is nearly level. Bog soils are also common along water courses, lake margins, thermokarst (alass-*Rus*) sites, and other depressions. These soils may have an organic thickness ranging from as little as 10 to 30 cm, but in some sites the organic accumulation may be as much as several tens of meters. The lower portion of Bog soils generally has large quantities of ground ice present.

Most higher land, especially that in the mountains, is much drier than that generally present in the lowlands. The tendency of freely drained soils to occur at higher elevations is brought about by a combination of factors—relief, stoniness, shattered bedrock, less massive ground ice, and others. On the plateaus and even the lowlands, freely drained soils may be present, but in such positions their areal distribution is small, usually being confined to dunes, terrace escarpments, kames, eskers, and similar situations where positive relief elements are present—and with those sites having a coarse-textured substrate with a dry permafrost condition.

In the well-drained sites of the arctic, soil formation follows evolutionary pathways somewhat similar to those of the boreal forest zone. At sites where favorable varieties of soil material coexist with proper relief elements, a thin humus layer forms. The upper mineral horizons develop a brown color, but usually there is little authigenic clay formation. Leaching in the well-drained arctic sites is, in effect, quite variable. Generally the upper horizons are slightly to moderately acid, pH values increasing with depth. On exposed positions, such as spurs and windswept ridges, however, where evaporation is greatest, the soil may accumulate pedogenic carbonate throughout the solum. On the other hand, in the valleys and plains where there are constructional land forms consisting of coarse-grained rock, podzolic soil characteristics tend to be present, especially on such materials as quartzite and granite. On other well-drained sites, particularly unstable sand dunes and alluvium, the soil may retain its original raw mineral appearance. Soils of the well-drained sites have been designated as *Arctic brown* (Tedrow and Hill 1955) but they have also been designated as *Sod soils* or *Sod-bare-rock soils* by Soviet investigators. The terms *Arctic brown soil* and *Sod-bare-rock soil* are not fully equivalent, however, because the latter term has been used primarily in the high mountains of central Asia, where climatic conditions are strikingly continental and therefore not fully analogous to conditions of the arctic. Targulian (1971) used the term *Podbur* for the well-drained, nonpodzolic soil of the tundra. The term *Podbur* approximates *Cryptopodzolic* or *Tundra illuvial humus*, *Cryogenictaiga ferruginous*, *Subarctic brown forest*, *Brown wooded*, and *Brown melanized* soil used by other investigators. Where there is visible podzolization within the soil, the term *Podzolic-Al-Fe-humus* soil has been used (Targulian 1971). Other similar terms, such as *Dwarf podzol* and *Miniature podzol*, have also been used. These terms provide for the most advanced type of soil formation on the well-drained sites of the arctic. They have also been used prominently in the subarctic of Scandinavia.

In addition to those previously mentioned, the following soils are present in special situations (Tedrow 1977):

Rendzina—shallow carbonate soils on limestone.
Ranker—'carpet soils,' organic mats over loose rock.
Shungite—black, carbonaceous shale soils.
Grumusol—gray, waxy, bentonite soils.
Volcanic ash soils (Andolike).
Lithosol and bedrock—shallow, rocky soils and exposed bedrock

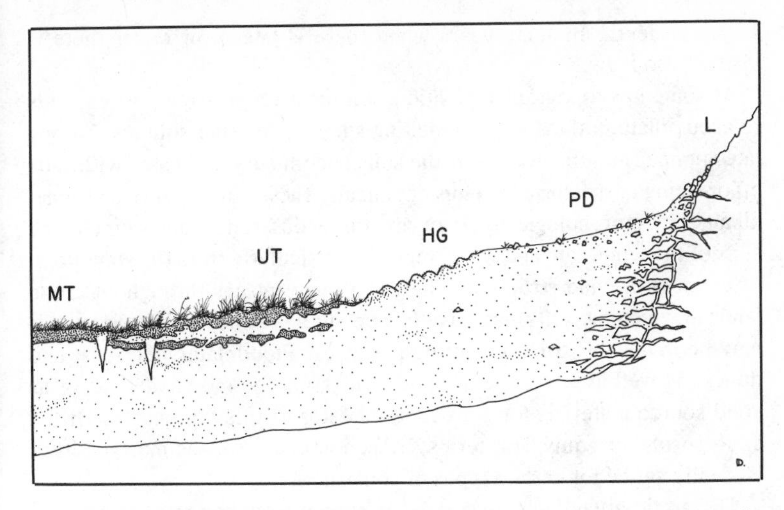

Fig. 6.4. Idealized cross section of the major soils of the polar desert (high arctic) zone: MT, Meadow tundra soil; UT, Upland tundra soil; HG, Soils of the hummocky ground; PD, Polar desert soil; and L, Lithosol.

Regosolic material—generally sands and gravel with little soil development. Recent dunes, fluvial deposits, talus, and other raw earthy material.

The high arctic

Within the high arctic (polar desert zone), the pattern of soil formation is somewhat different from that of the main arctic to the south. Figure 6.4 depicts a soil–landscape relationship of the high arctic. Summer ambient temperatures and precipitation values are lower than those farther south. Accordingly, large land areas have only a scattering of vascular plants, especially in the uplands. These barren conditions have been designated *polar desert, arctic desert, rock desert,* and *fell field,* among others. The well-drained sites may have a desertlike type of soil formation with a dry permafrost condition beginning at a depth of as much as 1 m. Nevertheless, this is a sufficient depth for the solum to develop without a significant influence from the permafrost. Polar desert soil generally has a desert pavement together with other desertlike features. Tundra soils are present in the high arctic, as is the case farther south, but profiles in the former are generally shallower, with the frost table being at about 25 to

35 cm in depth. In some years when there is late snow cover there is virtually no thaw.

At some low topographic positions within the high arctic, water tends to accumulate in the soil from melting snows and other sources. During late summer months, however, the soils become dry and hard, with salts efflorescing at the surface. Thus chemically these soils qualify as saline-alkali, but morphologically they are quite different from saline-alkali soils of the temperate and warm regions, particularly from the standpoint of temperature and structure. There are many places throughout arctic lands in which the soils may have few developed features, but these soils may be invaluable for engineering purposes. Floodplains, terraces, and dunes, as well as some rocky sectors of the arctic mountains, serve as good source materials for construction. Sites of this type may be listed as Regosols, or equivalent terms, by pedologists, but the material itself is ideally suited for certain types of construction.

The arctic areas have always been known to experience active solifluction, especially on the steeper land (Jahn 1961). The downslope movement of the soil can be recognized by such surface markings as lobes and terracelike forms. Where solifluction is quite active, it is generally best to designate the site as a solifluction area with notations as to the edaphic conditions of substrate, but without trying to ally it with a specific genetic soil.

Arctic pedologic soil classification schemes

Unlike most other climatic regions of the globe, the arctic has received only fragmentary consideration with regard to pedologic soil classification problems. At the present time the various proposed classification schemes should be considered rather tentative. Proposals by various governments, as well as some independent investigators follow.

U.S.S.R. classification of arctic soils

Although Soviet investigators have been engaged in pedologic arctic soil classification for many years, no one specific system has achieved recognition—especially in the treatment of lower taxonomic units of the arctic (Table 6.1). In principle, however, most Soviet investigators tend to base their work on the Dokuchaevian-derived concept, which considers the principles of zonality on a background of at least three main features: (i) composition and properties of the soil, (ii) ecologic condi-

TABLE 6.1.

Classification of arctic soils according to Soviet investigators

	U.S.S.R. (Ivanova 1956)	Soil classification (Makeev 1978)[a]
Arctic soils (polar desert zone plus transitional lands between polar desert and tundra)	Arctic gleyturf soils	1. Arctic desert
	Arctic cryptogley polygonal soils	2. Arctic typical humus
		3. Boggy cryogenic
		4. Arctic tundra raw humus gleyish
Tundra soils (tundra zone)	Tundra gleyey soils	5. Tundra raw humus gley
	Tundra gleyey podozolized soils	6. Tundra weakly peaty and peaty gley
		7. Tundra soddy gley
		8. Arctic tundra humus gleyed
		9. Arctic tundra humus alluvial
		10. Typical tundra gley
		11. Tundra humus illuvial

[a] Makeev recognizes arctic soil zonality in his scheme of classification.

tions of development, and (iii) dynamic processes of development. The classification schemes of Ivanova (1956) and Makeev (1978) give certain orders of classification, but they also consider lower orders of taxonomy.

U.S. Department of Agriculture classification of arctic soils

For years there were virtually no arctic pedologic investigations from the United States. With publication of *Soil Taxonomy,* by the U.S. Department of Agriculture Soil Survey Staff (1975), there was some attention given to arctic soils, although the discussions were confined largely to theoretical consideration plus renaming of soils described previously in the literature (Table 6.2).

The above work of the U.S. Department of Agriculture largely ignores the problem of soil zonation within arctic regions. The system is based largely on measurable properties, such as base saturation and color, plus certain other approximated or projected properties of the soil. The classification system of the U.S. Department of Agriculture has not been tested to any extent in the arctic areas except for a limited amount in Alaska.

Canadian Department of Agriculture classification of arctic soils

The classification system of the Canadian Department of Agriculture is similar in some respects to that of the U.S. Department of Agriculture, especially as applied to the arctic areas (Clayton et al. 1977). In principle, most soil names used in the northern fringes of the forested lands are extended into the arctic region with the prefix *cry* (cold) added to the stem of the term to indicate that the soil exists in a low temperature regime. The major arctic soils recognized at the subgroup level by Canadian investigators are given in Table 6.3.

With the above nomenclature the terms Podzol and Brunisol provide for the mature, well-drained sites, the Gleysol for the Tundra soils, and the Fibrisol, Mesisol, and Humisol for the various organic deposits. The Polar desert soil as described by Tedrow (1966) is included as a part of the saline Cryic Regosol.

Other proposals for arctic pedologic soil classification

The proposed classification scheme of Tedrow (1977) recognizes Great Soil Groups within the various arctic zones but does not restrict the

TABLE 6.2.
U.S. Department of Agriculture soil classification system for arctic areas

Order	Suborder	Great group	Formative elements
Entisols	Aquents	Cryaquents[1]	Cold-wet-recent soil
	Orthents	Cryorthents	Cold-main-recent soil
	Psamments	Crypsamments	Cold-sandy-recent soil
Inceptisols	Aquepts	Cryaquepts	Cold-wet-initial soil
	Andepts	Cryandepts	Cold-ando initial soil
	Umbrepts	Cryumbrepts	Cold-shade initial soil
	Ochrepts	Cryochrepts	Cold-pale-initial soil
Mollisols	Aquolls	Cryaquolls	Cold-wet-soft soil
	Borolls	Cryoborolls	Cold-northern-soft soil
Spodosols	Aquods	Cryaquods	Cold-wet-ashy soil
	Orthods	Cryorthods	Cold-true-ashy soil
	Humods	Cryohumods	Cold-humus-ashy soil
Histosols	Fibrists	Cryofibrists	Cold-fibrous soil
	Folists	Cryofolists	Cold-leafy soil
	Hemists	Cryohemists	Cold-partially decomposed soil
	Saprists	Cryosaprists	Cold-highly decomposed soil

[1] *Cry* is added to the term to form Cryaquent (cold-wet-recent).
Source: *Soil Taxonomy* 1975.

TABLE 6.3.
Classification of arctic soils according to Canadian investigators

Order	Great group	Subgroup
Podzolic	Humo-Ferric Podzol	Cryic Humo-Ferric Podzol
Brunisolic	Eutric Brunisol	Cryic Eutric Brunisol
	Dystic Brunisol	Cryic Dystric Brunisol
Regosolic	Regosol	Cryic Regosol
Gleysolic	Humic Gleysol	Cryic Humic Gleysol
	Gleysol	Cryic Gleysol
Organic	Fibrisol	Cryic Fibrisol
	Mesisol	Cryic Mesisol
	Humisol	Cryic Humisol

Source: Clayton et al. 1977.

TABLE 6.4.
Schematic diagram for classifying arctic soils

First order (zone)	Second order (Great Soil Group)	Third order (series)
Tundra soil zone	Well-drained soils Arctic brown soil Podzollike soil Mineral gley soils Upland tundra soil Meadow tundra soil Organic soils Bog soil Other soils Ranker soil Rendzina soil Shungite soil Grumusol Lithosol Regosol Soils of the solifluction slopes	Separations based on textural and mineral properties of the parent material, etc.
Subpolar desert soil zone	Well-drained soils Polar desert soil Arctic brown soil Mineral gley soils Upland tundra soil Meadow tundra soil	Separations based on textural and mineral properties of the parent material, etc.

First order (zone)	Second order (Great Soil Group)	Third order (series)
	Soils of the hummocky ground	
	Soils of the polar desert —tundra interjacence	
	Organic soils	
	Bog soil	
	Other soils	
	Regosol	
	Lithosol	
	Soils of the solifluction slopes	
Polar desert soil zone	Well-drained soils	Separations based on textural and mineral properties of the parent material, etc.
	Polar desert soil	
	Arctic brown soil	
	Mineral gley soils	
	Upland tundra soil	
	Meadow tundra soil	
	Soils of the hummocky ground	
	Soils of the polar desert —tundra interjacence	
	Organic soils	
	Bog soil	
	Other soils	
	Regosol	
	Lithosol	
	Soils of the solifluction slopes (may be a form of gley soil but usually well-drained)	

Source: Tedrow 1977.

Great Soil Group entirely to any one soil zone. Each Great Soil Group is then subdivided according to series (U.S.) (Table 6.4).

Fedoroff (1966), also working independently, proposed a classification scheme which includes the genetic soil variety plus the patterned ground forms. The system, although rather complex, should receive serious attention. It is based on genetic principles and includes both the soil variety and the associated patterned ground form. Fedoroff (1966) grouped the polar soils into the following categories:

Raw soils (Lithosols and Regosols, etc.)
'Little-evolved' soils with various subdivisions according to ice segregation
'Browned' soils with subdivisions according to degrees of development and eutrophism
Podzolic soils with subdivision
Halomorphic soils with subdivision
Hydromorphic soils

Further separation of the above units is based on degree of development, parent material, ground ice pattern, type of patterned ground, and ancillary factors.

7. Soil mechanics

The term 'soil mechanics' is used to designate that science which deals with the character and behavior of soils for engineering purposes. In contrast with agriculture, which is primarily concerned with the uppermost layers of soil adapted to the support of plant life, engineering and soil mechanics consider soils to include all earth materials, organic and inorganic, occurring in the zone overlying bedrock and composed of solid particles, void space, and water or gas or both. Mechanical rather than agricultural properties are of primary importance. Nevertheless, the science of soil mechanics does not involve merely the mechanical behavior of collections of inert solid particles. It also involves such intimately interrelated factors contributing to the mechanical characteristics of soils as the hydraulics and chemistry of the pore fluids and the physics, chemistry, and structure of the individual soil particles. The permeability of the soil to water is often a major consideration, particularly to the soil mechanics engineer involved in the design of water-control facilities, such as earth dams, cofferdams, or subsurface drainage or dewatering systems. Chemical characteristics of the soil may be especially important when steel piles, pipelines, or other features might be rapidly corroded if placed in contact with the soil. The thermal properties of soils are of fundamental importance when heat flow into, out of, or within the ground must be calculated or controlled for engineering purposes.

The most important nonfrost soil mechanics problems are those involving (a) stability and (b) *settlement* or *swelling*.

Stability failures in soils may occur as shear failures within soil masses, as in highway cuts and embankments, steep hillsides, or earth dams, and under footings which support structures, or under vehicle or aircraft wheel loads. Shear failures may also occur at the contact surface between soil and another material, as on the below-ground surfaces of a friction-type foundation pile. Other common types of stability failure are sliding or overturning of retaining walls and collapse of inadequately braced construction trenches. Shear failures may occur by the sliding of entire masses of soil along well-defined surfaces, by creep, by solifluction, and by liquefaction and flow. These modes are further discussed in relation to terrain evaluation in permafrost areas in Chapter 16. Flow of water through soil can significantly affect stability because of the drag or friction force which seepage produces in the direction of motion of the water. Under certain conditions of upward flow, the shear strength of granular soils may be reduced to zero. Seepage emergence or flow of water on the ground surface may cause erosion, involving the grain-by-grain detachment and movement of soil particles. One of the most useful concepts in soil mechanics is that of protective filters which permit seepage flow to occur without surface erosion, movement of particles or development of piping within the soil, or occurrence of a quick condition.

Settlement problems may occur when the soil is subjected to compression under load. The process by which compression occurs gradually at constant load, with forcing of soil particles closer together, decrease of water content, and transfer of stress from the water to the solid particle structure is termed *consolidation*. The water in the voids carries part of the stress until consolidation is complete, yet does not contribute to the soil's resistance to shearing; thus saturated, compressible, fine-grained soils may experience shear failure if loaded too rapidly in relation to the rate at which fluid can escape from the voids during the consolidation process. On the other hand, certain clay soils tend to increase substantially in volume when allowed to absorb water, and swelling pressures which develop may be very high, with results similar to frost heave.

In case of either settlement or swelling, difficulties may arise both from displacements relative to adjacent facilities or structures and from differential displacements within the area of the foundation caused by variations in soil properties and loadings. The differential vertical movements may cause structural damage to buildings, including cracking of walls and floors, breaking of service pipes, and tilting or wrecking of the entire structure.

To cope with such problems, soil mechanics engineers rely, as a first step, on the systematic evaluation of the properties of the soils involved. Classification of the soils in accordance with engineering soil classification systems, as described in Chapter 12, gives the soil mechanics engineer an immediate understanding of the general engineering characteristics of the soil. This may sometimes be sufficient for the design of simple projects. For projects requiring more detailed information, additional tests may be performed to measure such specific properties as permeability, compaction, shear strength, consolidation, and dynamic response characteristics. On some projects, full-scale field tests, such as pile driving and pile-load tests, may be necessary before design can be completed.

Using the soil data, theoretical analyses and procedures developed through accumulated practical experience are then applied in order to achieve practical design and construction solutions for the soils aspects of such features as foundations of buildings and other structures, retaining walls, pipelines, airfields, highways, drainage structures, and dams and embankments.

Because of the variability of soil in nature, the margin of possible error in solutions to most engineering problems involving soil mechanics is greater than it is for homogeneous materials, such as steel or other manufactured products. Much practical knowledge and experience, in addition to competence in scientific analysis and application of theory, are required to interpret the subsurface information successfully and render designs that are not only reliably safe and effective but also as economical as possible. Nevertheless, the development of the science of soil mechanics in recent decades has made possible the avoidance, on the one hand, of the gross overdesign and excessive use of safety factors which were once common and of the high frequency of failures, on the other.

8. Engineering characteristics of soils in cold regions

In seasonal frost and permafrost areas the soil mechanics engineer encounters problems not present in nonfrost areas. When soil freezes, mechanical, hydraulic, and other properties of the material are markedly altered. Ice segregation, frost heaving, shrinkage cracking, and subsequent thaw-settlement and thaw-weakening may occur in the seasonally frozen strata in both seasonal frost and permafrost areas. Ice formed in previous times by ice segregation may also be found within permafrost formations. In permafrost areas, more or less vertical ice wedges formed in a repetitive seasonal contraction-expansion process are found, together with various special frost-related surficial features. "Fossil" ice which has been preserved as a result of burial by a landslide or other event may also be encountered. The tendency of frozen soils to deform progressively under load requires special soil mechanics analytical procedures. When permanently frozen ground containing massive ground ice thaws, extremely damaging settlements may occur. The type of engineering application determines which of several soil properties is most important in a particular case. For example, shear strength, frost susceptibility and thaw-consolidation characteristics are most important for footing-type foundations.

Ice segregation

Ice segregation may be defined as the growth of ice within soils in excess of the amount which may be produced by the in-place conversion of the original void moisture to ice. The segregation occurs most commonly as distinct lenses, layers, and masses, commonly, but not always, oriented at right angles to the direction of heat loss. The excess ice may also occur in interstitial form, more or less uniformly distributed through the soil mass and often not readily apparent to the unaided eye. When ice segregation is occurring, water is strongly attracted to the position at which ice is forming from the unfrozen soil below the plane of freezing. The most obvious result of ice segregation is heaving of the ground surface, usually approximately equal in amount to the total thickness of ice strata formed.

In order for ice segregation to occur, three conditions must be present simultaneously (Linell 1960):

(i) Freezing temperatures must penetrate the soil.
(ii) The soil must be frost-susceptible.
(iii) A source of water must be available.

Penetration of freezing temperatures into the ground has been discussed in Chapter 4.

Frost susceptibility

The following criteria for frost-susceptible soils (Linell et al. 1963) are used by the U.S. Army Corps of Engineers for pavement design purposes; they are based on a criterion originally published by Casagrande (1931) and additional research by the Corps of Engineers:

> Most inorganic soils containing 3 percent or more of grains finer than 0.02 mm in diameter by weight are frost susceptible. Gravels, well-graded sands, and silty sands, especially those approaching the theoretical maximum density curve, which contain 1.5 to 3 percent finer by weight than 0.02 mm size should be considered as possibly frost-susceptible and should be subjected to a laboratory frost-susceptibility test to evaluate actual behavior during freezing. Uniform sandy soils may have as much as 10 percent of their grains finer than 0.02 mm by weight without being frost-susceptible. However, their tendency to occur interbedded with other soils usually makes it impractical to consider them separately.

Pore-size distribution or heaving pressure would be more fundamental indicators of frost susceptibility, but the grain size criterion has the advantage of being easy to measure.

Figure 8.1 shows the results of the Corps of Engineers' measurements of the relative frost-heaving qualities of soils, classified by the Unified Soil Classification System, when tested in the laboratory in an open system with relatively unlimited availability of water and under standardized conditions. The values of rate of heave shown in Fig. 8.1 are comparative only and do not represent rates of heave which will occur under specific or even average field conditions. As may be seen in Fig. 8.1, heave potential is not zero when the percentage of grains finer than 0.02 mm by weight is equal to or immediately below the values stated in the above paragraph. It is low enough, however, to be considered acceptable for most pavement applications. For design of foundations for structures or for other engineering applications these criteria may or may

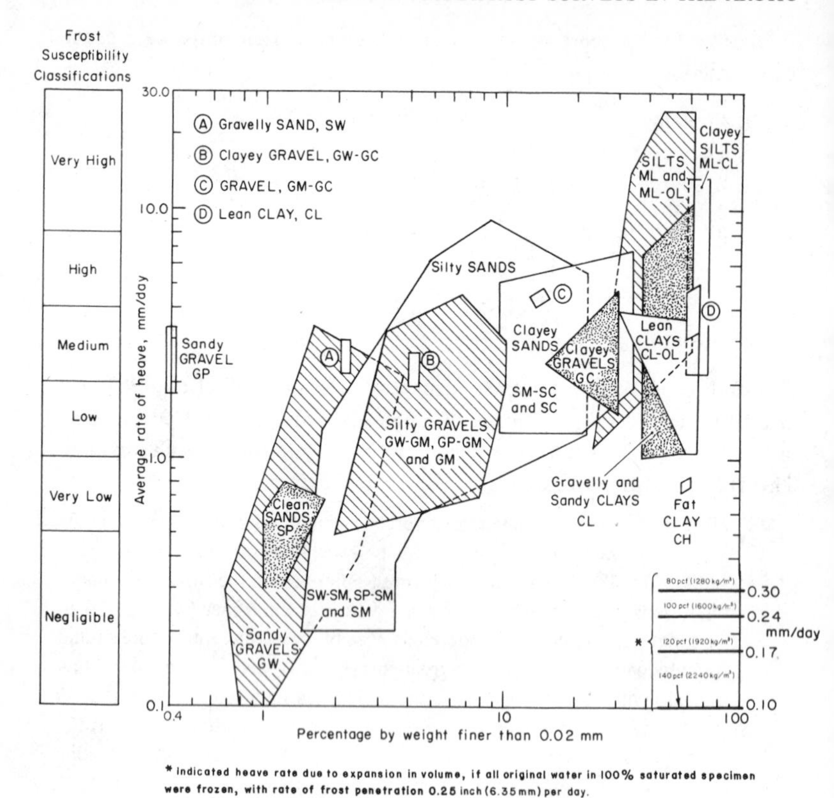

Fig. 8.1. Summary of average rate of heave versus percentage finer than 0.02-mm size for natural soil gradations (Kaplar 1974).

not be directly suitable, depending upon the nature and requirements of the particular construction. Nevertheless, they provide very useful rule-of-thumb guidance if used with proper discretion.

Clean GW, GP, SW, and SP gravels and sands with a negligible percentage of material finer than 0.02 mm are so completely nonheaving that frost heave and thaw-weakening in the annual frost layer may usually be disregarded as design factors. Fluffing of the surface of unconfined, relatively clean sands and gravels may occur on freezing, even though well-drained, but only a small degree of confinement, as by a pavement, normally eliminates this effect. If clean gravels and sands are 100 percent saturated, there will be a possibility of a small amount of expansion corresponding to the increase in volume that occurs when water changes to ice. Even this will not occur, however, if this expansion can be compensated for by escape of an equivalent volume of pore water into and through the soil below the plane of freezing.

Effect of moisture on frost heave

In fine-grained soils, ice segregation may be expected to occur on freezing if water is available. During freezing of frost-susceptible soil, water is drawn from the voids of still unfrozen soil to produce ice lenses or other forms of excess ice. Even when a closed system exists because of the presence of an impervious stratum or frozen layer at shallow depth, which prevents supply of additional water, ice segregation can occur by withdrawal of moisture from unfrozen soil above this stratum. If the supply of free water thus available is small, frost heave will necessarily be limited, and compression of the soil immediately underlying the freezing front by the moisture tension created by the withdrawal of moisture may cause the heave of the overlying ground surface to be even smaller or negligible. If the water drawn to the plane of freezing from the voids of the soil at and immediately below the plane of freezing can be easily replaced, however, as from an aquifer layer or water table at relatively shallow depth, heave can be very substantial. In most cases the heave of an overlying surface is usually approximately equal to the total thickness of the ice layers formed.

Perched water tables can supply water about as effectively as true water tables. Although lowering the water table in a seasonal frost area may be expected to reduce frost heave, the drainability by gravity of most frost-susceptible soils is limited, and the moisture retained in the soil voids after drainage may still be sufficient to produce unacceptable heave. Similarly, moisture available within the active layer at the end of summer in permafrost areas can be sufficient to produce very significant heave. Such heave may, in permafrost areas, involve greater uplift forces on structures than in seasonal frost areas because the lower soil temperatures in the more northerly areas make possible the development of higher frost grip stresses on foundation members.

Frost heave and stress

It has been shown under both laboratory and field conditions that increasing intergranular pressure normal to the freezing plane decreases

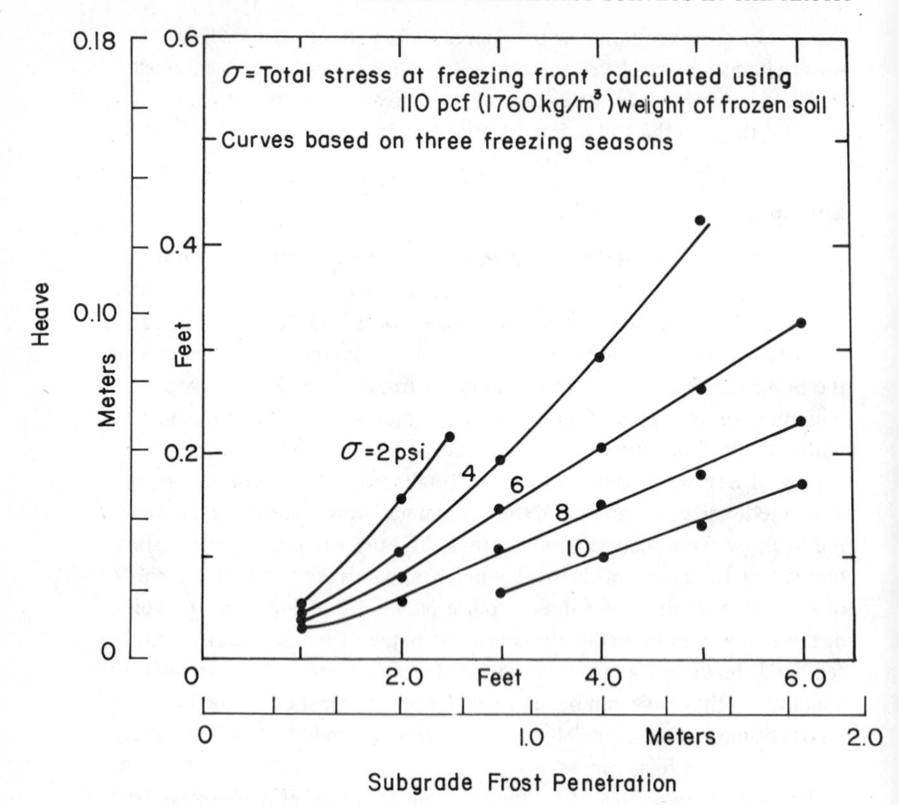

Fig. 8.2. Heave versus frost penetration for various total stresses, surcharge field experiment (Linell, Lobacz, et al. 1980, based on data from Aitken 1974).

the rate of frost heaving. Figure 8.2 shows, for example, the results of a field experiment on an active layer silt subgrade over permafrost near Fairbanks, Alaska. Seasonal heave is plotted versus subgrade frost penetration for various intensities of total pressure at the freezing interface. It may be seen, for example, that for subgrade frost penetration of 4.0 ft (1.22 m) the heave under 10 psi (69 kN/m^2) pressure was only about one-third of the heave under 4.0 psi (27.5 kN/m^2) pressure. The effect on heave of a given increase of pressure is less for clay soils than it is for silts.

Figure 8.3(a) shows, for a variety of soils, maximum frost heave pressures measured under conditions of essentially complete restraint, plotted against permeability of the unfrozen soil. This figure suggests that heave can be prevented if pressures at the freezing plane equal or exceed such values. This is considered a valid conclusion for the silts and coarser soils. As pointed out in Chapter 4, however, additional amounts of moisture freeze progressively in clay soils with decrease of temperature after initial crystallization, and, as reported by Keune and Hoekstra (1967), significant ice-lens growth can continue behind the freezing front because of moisture transport by unfrozen adsorbed moisture films. Such transport is much less important in silts and coarser soils than in clays. Ice formation at lenses behind the freezing front occurs at temperatures lower than the initial crystallization temperature. Keune and Hoekstra (1967) have shown that the maximum heave pressure that can be developed at these depressed temperatures increases as the temperature at the growing ice lens decreases, as illustrated in Fig. 8.3(b). Radd and Oertle (1973) have reported even higher pressures. Thus, the potential heaving pressures in clay may be more than might be concluded from Fig. 8.3(a), even though the potential amount of heave might be small. Furthermore, the single clay in Fig. 8.3(a), Lebanon Clay, is a relatively coarse-grained clay, as may be seen by comparison of its permeability with the table of permeability ranges in Table 8.2, presented later in this chapter, and it may be inferred from Fig. 8.3(a) that medium and fat clays could be even more striking in their pressure effects.

The data in Figs. 8.2 and 8.3 involve heave factors acting in compression against surfaces oriented at right angles to the direction of freezing. Figure 8.4 shows, for comparison, average tangential shear stresses generated by drag of the seasonally frozen layer on the lateral surfaces of piles restrained against vertical movement. Because these stresses are averages for a range of temperature values with depth within the frozen layer, they do not indicate the peak stresses. The curve for the 1962–3 test on 8-in. (20 cm) diameter steel pipe-pile shows especially well the surges of heave stress which were produced by waves of especially low winter temperatures, followed by relaxation.

Heave force effects

As illustrated in Fig. 8.5, for various structural situations, heave acts against the base of a slab of frozen soil rather than directly against foundations or structures. Because of the bending strength of the frozen soil layer, uplift force is derived from a much wider area of heave than the actual area of the foundation or structural unit in contact with the

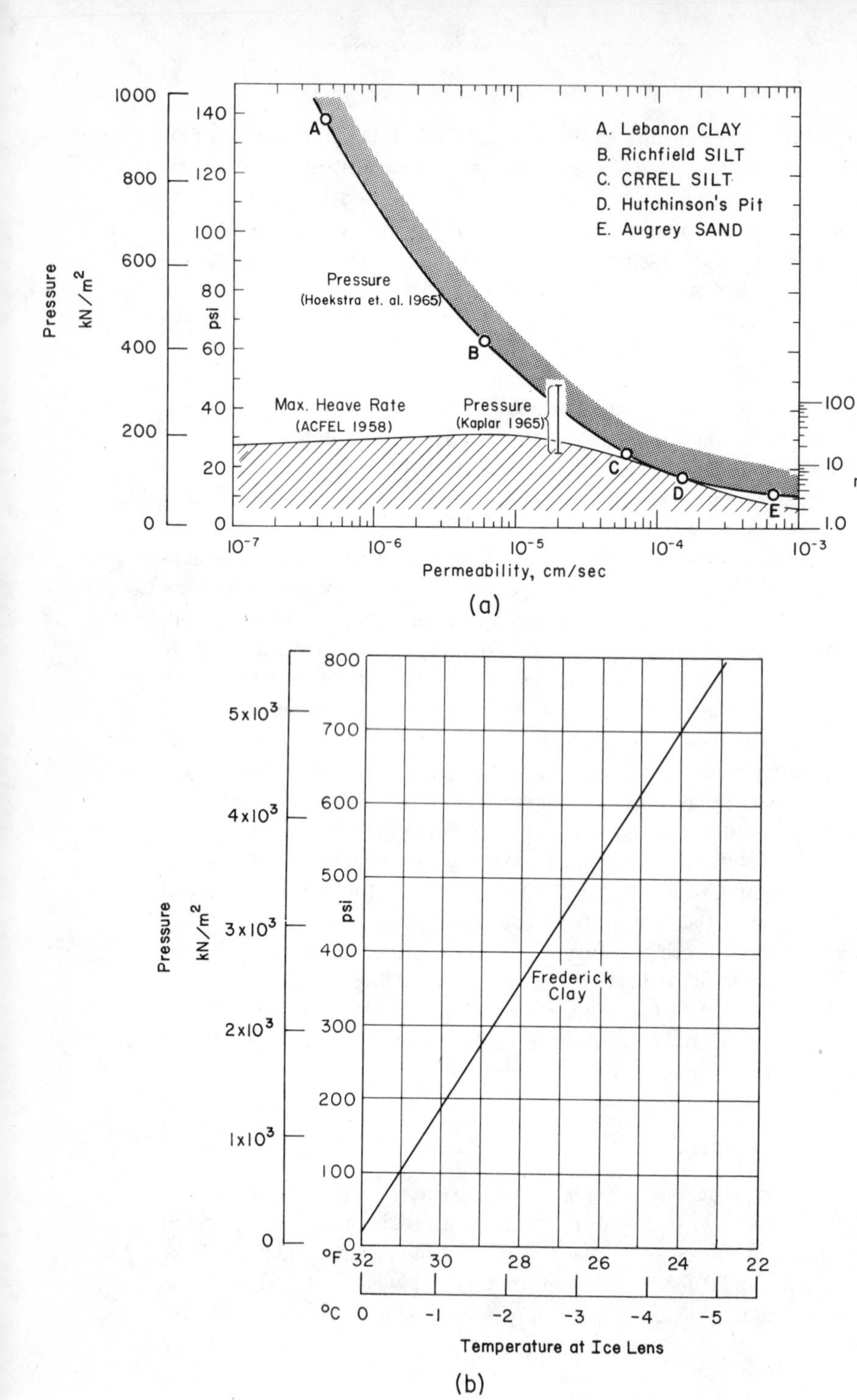

A. Lebanon CLAY
B. Richfield SILT
C. CRREL SILT
D. Hutchinson's Pit
E. Augrey SAND
Pressure
(Hoekstra et. al. 1965)
Max. Heave Rate
(ACFEL 1958)
Pressure
(Kaplar 1965)
Pressure
kN/m²
psi
Heave Rate mm/d
Permeability, cm/sec
(a)
Frederick Clay
Temperature at Ice Lens
(b)

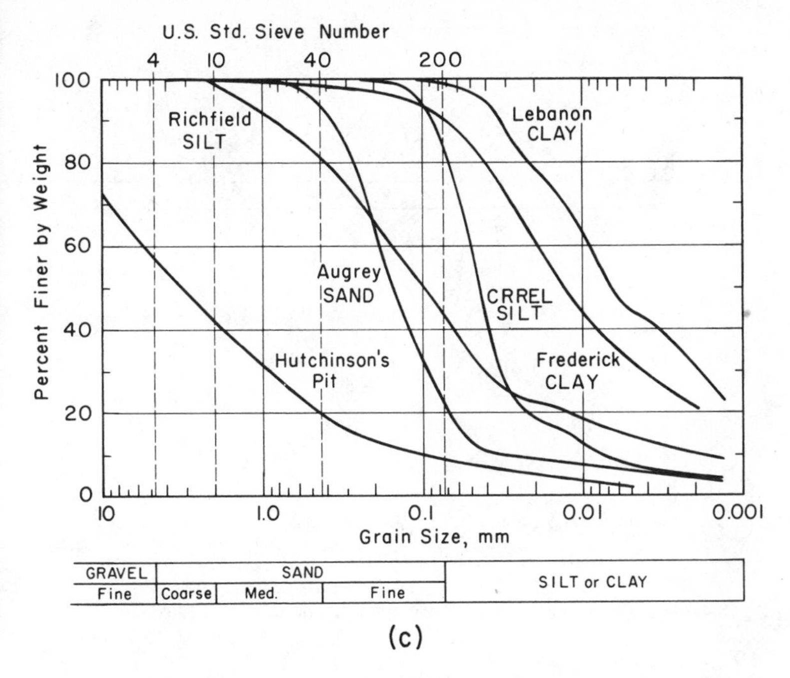

Fig. 8.3. Some reported maximum frost-heave pressures: (a) pressure and heave versus permeability; (b) temperature at growing ice lens versus maximum pressure developed at that depth (Keune and Hoekstra 1967); (c) grain-size distribution of soils (Linell, Lobacz, et al. 1980).

soil, greatly complicating estimates of heave force. Where the ground slopes, or is supported by a retaining wall, the heave does not act vertically but in the direction of heat flow. Drag on lateral surfaces of foundation elements embedded in the seasonal frost layer not only exerts uplift forces but also tends to reduce heave of the soil immediately in contact.

Compression, consolidation, and settlement

Unfrozen soils tend to decrease in volume, or be compressed, when subjected to pressure. When the voids are filled with water, the rate at which this volume change can occur tends to be restricted because of the time required to drain out excess water equal in amount to the volume

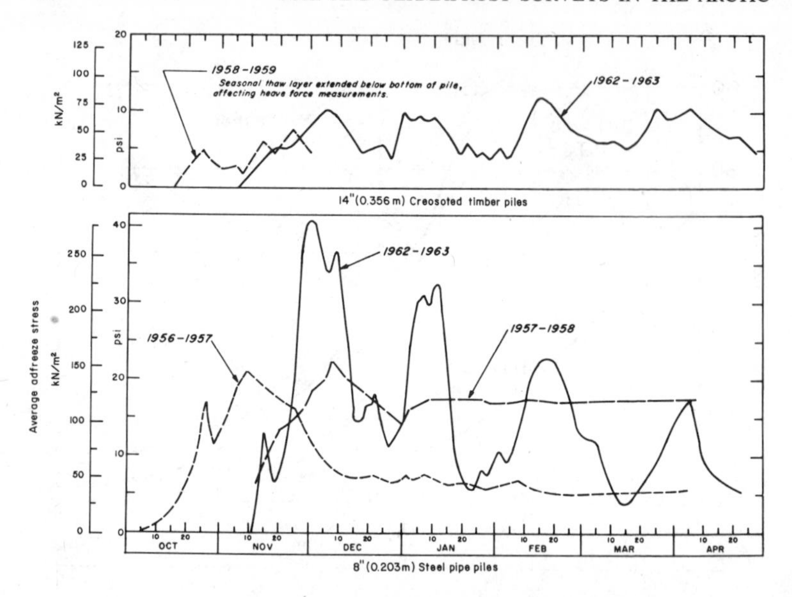

Fig. 8.4. Average adfreeze stress versus time (Crory and Reed 1965).

change. The process of gradual expulsion of water from soil under a constant compressive load is termed *consolidation*. Methods for evaluating compression and consolidation characteristics of unfrozen soils are described in many standard reference publications, such as the book by Terzaghi and Peck (1967).

For ordinary engineering purposes, gravels, sands, and silts of substantial moisture content may be assumed to have zero potentials for compression and consolidation when frozen. Frozen clays may experience some compression and consolidation under load, depending on the percentage of soil moisture that is frozen. When any soils containing excess ice are thawed, however, a special consolidation situation develops. The water which is in excess of the amount that can be accommodated in the voids of thawing soil can not drain downward as long as there is impervious frozen soil below. If lateral drainage is restricted or too slow, the moisture is forced to drain upward toward the surface, tending to cause very wet terrain conditions and saturation of pavement-supporting courses, the excess water sometimes emerging through cracks and joints in the pavement. The combination of lateral and upward thaw drainage that occurs under a pavement is illustrated in Fig.

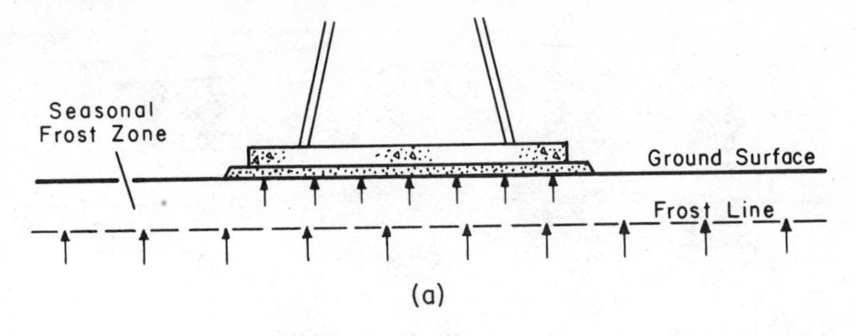

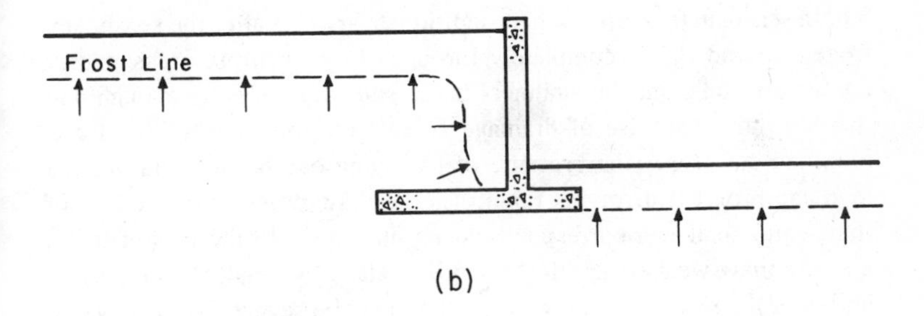

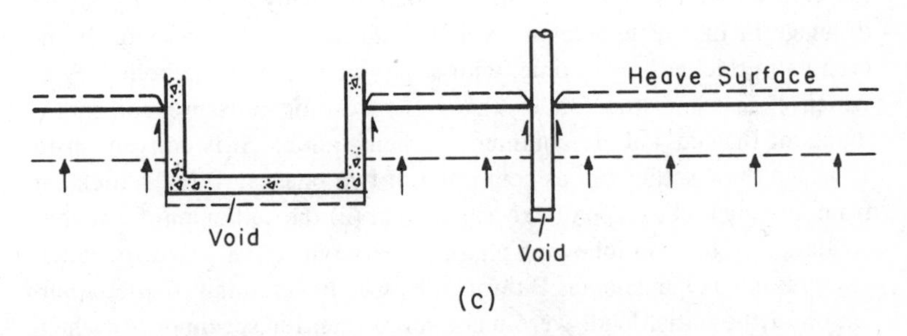

Fig. 8.5. Heave force effects: (a) heaving of soil in seasonal frost zone causing direct upward thrust on overlying structural elements; (b) freezing of frost-susceptible soil behind walls causing thrust perpendicular to freeze front; (c) force at base of freezing interface tends to lift entire frozen slab, applying jacking forces to lateral surfaces of embedded structures, creating voids underneath. Structures may not return to original position on thawing (Linell, Lobacz, et al. 1980).

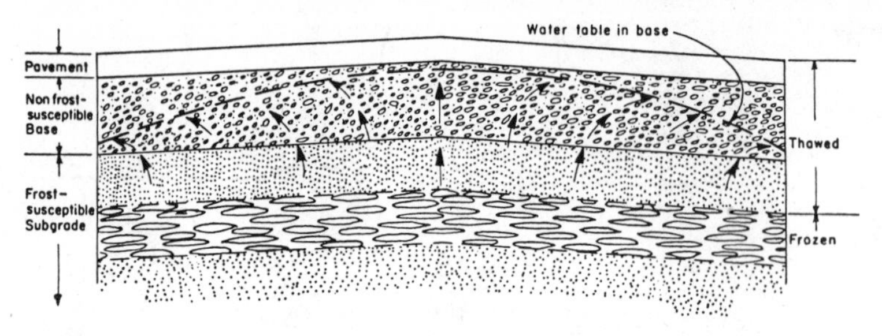

Fig. 8.6. Moisture movement upward into pavement base course during thaw (U.S. Army/U.S. Air Force 1966d).

8.6. In seasonal frost areas, this condition disappears after the seasonally frozen ground thaws completely through. In permafrost areas it may continue throughout the summer. If the soils are pervious enough and there is sufficient ease of drainage so that consolidation of the frost-loosened soil under the pressure of overlying overburden and applied load can progress nearly as rapidly as the thaw progresses, the rate of thaw settlement will correspond closely in time with the rate of thaw, and the thaw-weakening effects may be relatively small. If the permeability of the soil is low, thaw-weakening may be severe in intensity and duration.

Thaw-consolidation of frozen soil containing excess ice differs from the consolidation of unfrozen soil as conventionally analyzed for single drainage in that it involves a layer of material which continuously increases in thickness with time, with a moving nondrainage boundary at which excess moisture under excess hydrostatic pressure and increments of thawed soil are continuously being added. It is convenient to visualize thaw settlement as composed of two phases: (i) the settlement from melting and escape of excess ice, and (ii) the settlement from consolidation of the soil following phase (i). Note in Fig. 8.7 that the initial compressions of specimens B through E, which were thawed after application of the initial load, were much more than for specimen A, which was thawed prior to application of the initial load.

Figure 8.8 illustrates the effects of thawing and subsequent reconsolidation on the bearing capacity of flexible and rigid pavements over frost-susceptible subgrade soil as measured by the amount of load required to produce 0.1 in. (2.54 mm) deflection in plate bearing tests. These measurements were made in a seasonal frost area, but the same

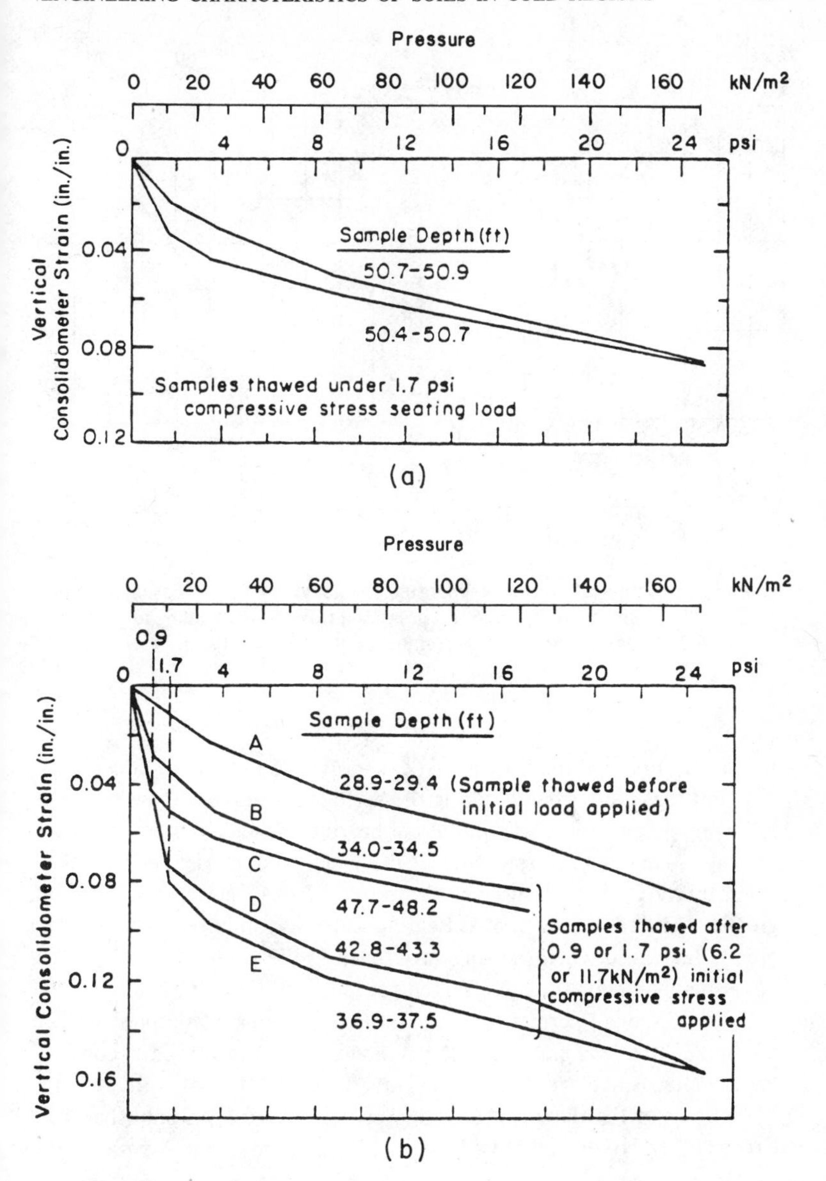

Fig. 8.7. Consolidation test results for undisturbed samples from two drill holes at the same site. Tests were performed using fixed-ring consolidometer. Specimen diameter 2.5 in. (64mm), height 0.8 in. (20 mm). (Linell, Lobacz, et al. 1980.)

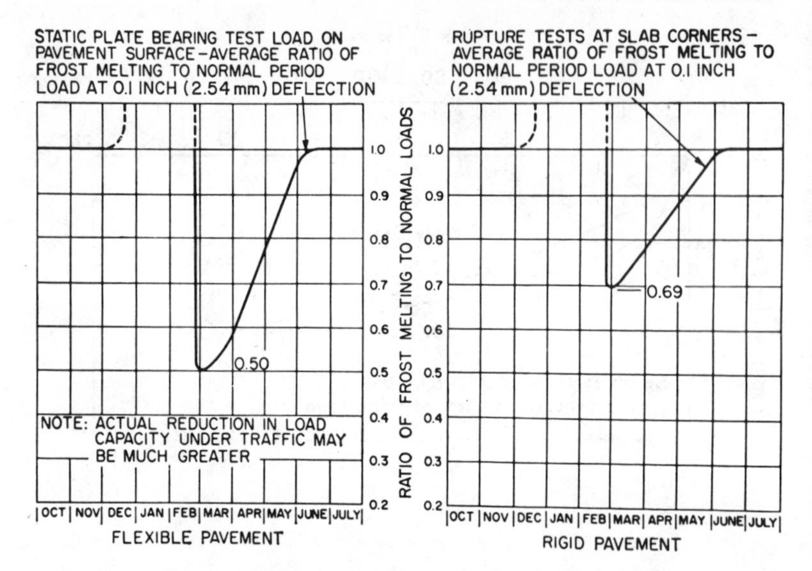

Fig. 8.8. Reduction in pavement supporting capacity. Curves are based on static load tests from nine flexible paved test areas and from corner loading tests on six rigid paved test areas at six locations (after Linell and Haley 1952).

pattern occurs in permafrost areas, except that the period of consolidation and regain of strength during the spring and summer thaw season, following the precipitous initial loss of bearing capacity at start of thaw, extends to the start of freeze-up at the end of summer. Similarly, the trafficability of natural surfaces composed of frost-susceptible soils is greatly reduced during thaw. Bearing capacity and trafficability are greatly increased when the soils are frozen in the winter season. Snow cover may delay or sometimes even prevent refreezing, however.

If degradation of permafrost occurs, the thaw-consolidation process is not limited to a seasonal effect. It can then go on for many years, continuing as long as thaw continues to occur.

Analyses of the thaw-consolidation problem have been published by Crory (1973), Nixon and Morgenstern (1973), and others.

Strength properties

Strength is the ability to resist mechanical forces. In soil mechanics the capacity of soil to resist shear, or sliding between internal surfaces in a soil mass, is a critical soil property. Shearing strength of soil may often be approximated by the Coulomb equation:

$$s = c + p \tan \theta$$

where

s = unit shearing strength
c = unit cohesion
p = unit effective normal pressure on a surface
θ = friction angle of the soil
$\tan \theta$ = coefficient of internal friction.

The cohesion component of the strength, c, is the internal cohesive strength of the soil which provides resistance to shearing even if there is no pressure acting normal to the plane of sliding. The friction component, $p \tan \theta$, of the strength is supplied by the grain-to-grain contact and interlocking of soil particles; when pressure normal to the sliding is increased, the friction component increases. The value of θ usually varies somewhat with the normal force p, but assumption of a fixed value for a given soil, as in the above equation, is often an acceptable simplification. The effective normal pressure, p, in the solid phase of unfrozen soils may vary with changes in the porewater pressure caused by seepage, consolidation, or volumetric changes during shear. The volume change during shear depends on such factors as the normal or confining pressure, the degree of saturation, the permeability of the soil, the rate of strain, the pressure or stress history, and the degree of compaction or relative densification. If fully saturated, relatively impervious soils are sheared rapidly, the actual volume change may be negligible, but the tendency to change is reflected in corresponding changes in pore pressure, thereby affecting strength. Unfrozen soils vary in their shear-strength characteristics from some clays whose shearing resistance may be assumed to be governed solely by cohesion ($\theta = 0$) to cohesionless granular soils whose shearing resistance is governed solely by the friction component.

The strength characteristics of soils may be represented in a *Mohr diagram,* which provides a graphical representation of the relationship between principal stresses at a point and the normal and shearing stresses at the same point on planes inclined with the planes of the principal stresses. In the Mohr diagram the abscissa represents the normal stress, and the ordinate represents the shearing stress. Figure 8.9 shows a Mohr diagram for frozen Manchester fine sand for tests performed on cylindrical specimens without confining pressure. For each test condition, such as at 20°F (−6.7°C), one semicircle has been drawn

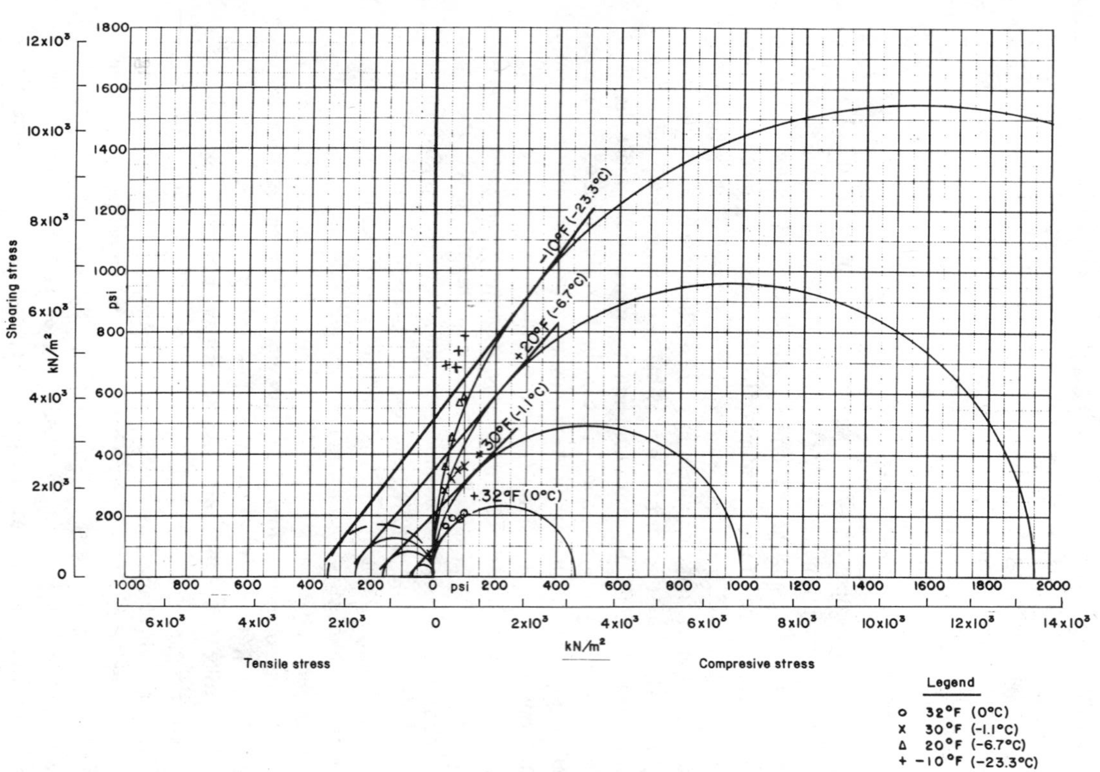

Fig. 8.9. Mohr diagram of stress conditions at failure, frozen Manchester fine sand. Plotted points indicate the average maximum shear stress obtained from direct shear tests. See Fig. 8.12 for soil gradation (SNH). (Kaplar 1953.)

on the figure of radius equal to half the unconfined compressive strength, the left end of which is at the origin of the diagram, and another of radius equal to half the tensile strength, the right end of which is also at the origin. A few average maximum shear stress values obtained by direct shear tests have also been plotted on Fig. 8.9. A curve drawn tangent to the test circles represents stress conditions at failure. Such curves are known as *Mohr envelopes*. Since only two Mohr circles are available on Fig. 8.9 for each test temperature, the Mohr envelopes have been drawn as straight lines. Nevertheless, the positions of the direct shear test points for 100 psi (690 kN/m^2) normal stress on Fig. 8.9 suggest that the relationships are probably curvilinear. The position and shape of the Mohr envelopes can be further developed if triaxial tests are performed to provide additional Mohr circles. To prepare a Mohr circle from triaxial compression test data, the lateral stress and maximum principal stress values at failure are plotted as the left and right abscissa of the circle, respectively. The straight line Mohr envelope is a plot of the previously described Coulomb equation $s = c + p \tan \theta$. The shear stress value at the intersection of the line with the vertical axis is the cohesion, c. The angle which the line makes with the horizontal is the angle of internal friction, θ.

Strength properties of frozen soils depend on such variables as gradation, density, degree of saturation, ice content or percentage of moisture frozen, temperature, dissolved solids, and rate of loading. When soil freezes, the ice formed in the voids acts as a cementing agent and void filler, affects the soil capacity for volume change during strain, and may contribute directly to strength properties in various degrees. Because the ice and soil fractions usually reach their maximum strengths at different degrees of strain, their maximum strengths tend not to be additive. For example, in cohesionless, granular soils which retain grain contact and grain interlock during freezing, the ice matrix contributes to strength at very low strains, but at high strains the ice matrix has already failed, and the ultimate strength becomes a function of the soil friction component (Sayles 1973). In soils containing substantial excess ice, on the other hand, the ultimate response characteristics may be controlled by the ice component of the material rather than the soil fraction. In frozen clays, strength is determined by a complex combination of the effects of soil cohesion, soil friction, ice content and distribution, and unfrozen moisture.

Figure 8.10 shows variations of shear strength with temperature for a variety of soils, as determined on specimens frozen in the laboratory and

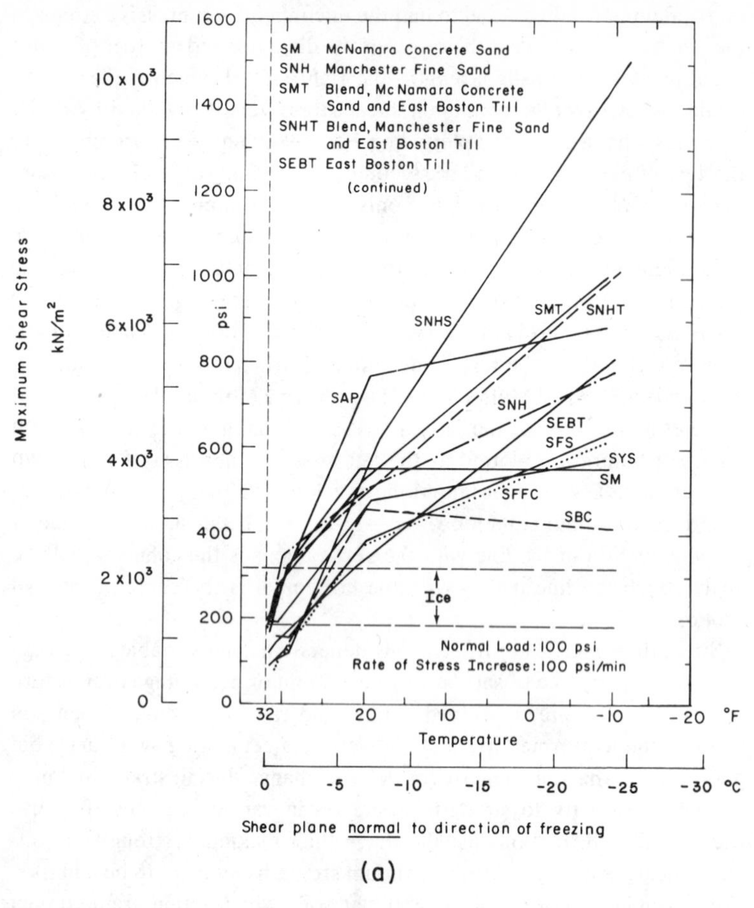

Fig. 8.10. Summary of maximum shear stress of frozen soils vs. temperature. See Fig. 8.12 for soil gradations. Average values for fresh water ice fell between the values shown (Kaplar 1954).

tested in direct shear under 100 psi/minute (690 kN/m^2/min) rate of stress increase and 100 psi (690 kN/m^2) normal stress. Results are shown both for tests in which the shear plane was oriented normal to the direction of freezing and for tests in which the shear plane was parallel to the direction of freezing. The highest shear strength values were obtained on the frozen New Hampshire silt with the shear plane oriented normal to the direction of freezing. At low temperatures, the lowest values were obtained on the frozen Boston blue clay. Tests were also made on speci-

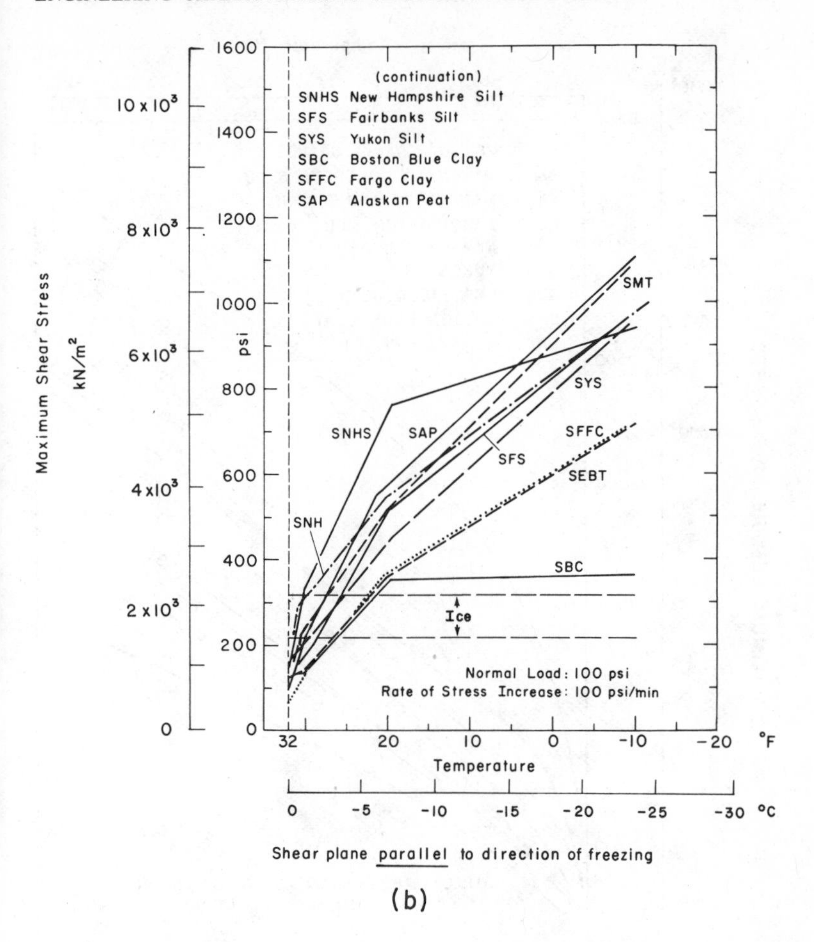

(b)

mens of fresh water ice frozen in the laboratory, under the same conditions as for the soil specimens. High and low limits for the average ice test values are shown on Fig. 8.10 in lieu of test curves because of the limited numbers of ice tests performed at each temperature level under the 100 psi (690 kN/m^2) normal load and 100 psi/minute (690 kN/m^2/min) rate of stress increase for which the figure has been prepared, and the scatter of these values. However, the results of the entire program of direct shear tests on ice, involving a total of 144 tests using four normal pressures, four rates of stress increase, four temperature conditions, and two orientations of the specimens with respect to the direction

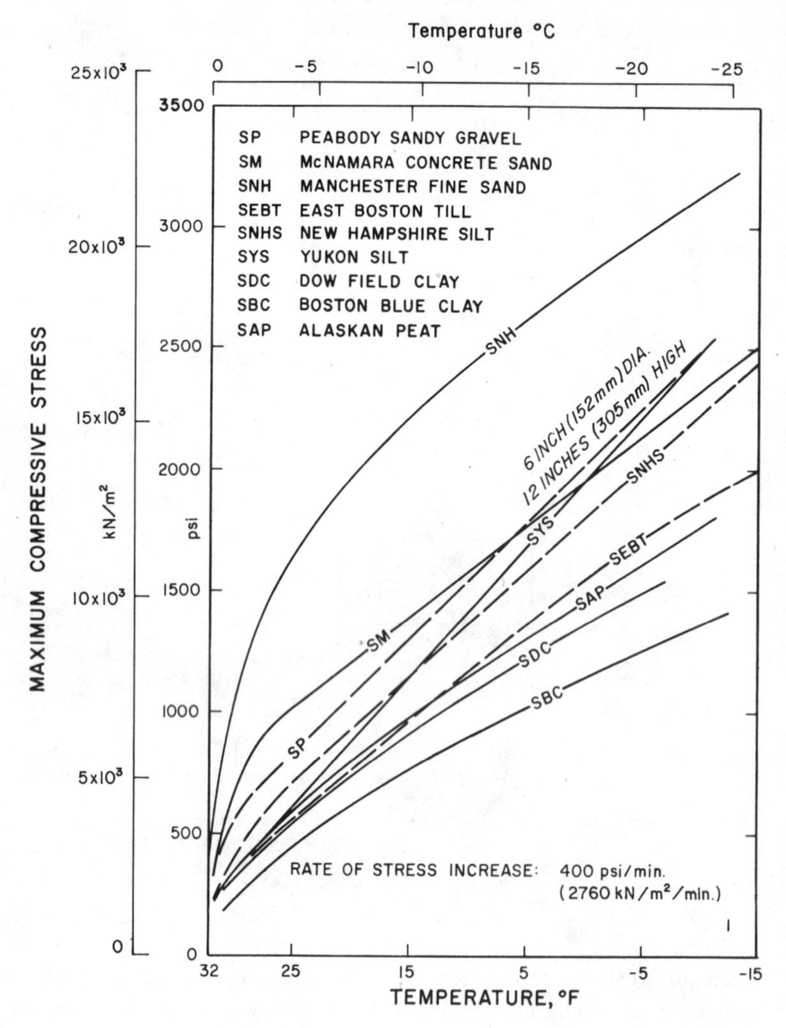

Fig. 8.11. Summary of maximum stress in compression versus temperature. See Fig. 8.12 for soil gradations (Kaplar 1954).

of freezing, showed a consistent failure of the ice to gain shear strength with decrease in temperature. Other investigators have reported similar results. At and just below the freezing point the ice was stronger in shear than the majority of the frozen soils at the same temperature, although at lower temperatures it was much weaker. Because of the effect of normal load, the magnitude of the ice shear strength values shown on Fig. 8.10 are higher than those reported by other investigators who did not use normal load.

Figure 8.11 shows compressive strength versus temperature for these same materials, and Fig. 8.12 shows gradation curves for the soils of Figures 8.9 through 8.11. Note that none of the soils as tested contained particles larger than ¾ in. (19 mm) except Peabody sandy gravel, and possible effects of cobbles and boulders are not represented.

Soils frozen at a high degree of saturation with ample moisture available characteristically exhibit creep, or continuing slow deformation under constant stress, at as low as 5 to 10 percent of the rupture strength measured in rapid loading, as illustrated in Fig. 8.13. They also exhibit lower strengths as rates of loading are decreased or the period of time that the soil will be subjected to a fixed stress is increased, as shown by Fig. 8.14, which presents curves of failure stress in compression versus temperature for frozen Manchester fine sand for various periods to failure, together with a curve of stress levels that can be sustained indefinitely. Gradation curves for the materials in both Figs. 8.13 and 8.14 are presented on Fig. 8.12. As demonstrated by Figs. 8.13 and 8.14, allowable working stresses in frozen soils are usually much lower than might be expected from conventional short-term loading tests, and, when creep effects may be significant, special engineering design approaches are required. Settlement of a foundation by creep at the rate of 1/100 in. (0.254 mm) per day would accumulate to 3.65 in. (93 mm) per year or about 3 ft (0.915 m) in 10 years. Even a relatively coarse-grained soil which is highly stable under nonfrost conditions may exhibit such behavior if ice is present in sufficient amount and form. Soils frozen at substantially less than 100 percent saturation, however, such that grain-to-grain contact and interlock are maintained, may not exhibit these strength deficiencies.

The tendency of soil to adhere to surfaces with which it is in contact when it freezes can be both destructive, as when foundation members are gripped and lifted by frost heave, or constructive, as when piles are designed to be supported in permafrost and derive their load-supporting capacity primarily from adfreeze tangential shear strength. Fig. 8.15 shows typical ultimate and sustainable average adfreeze bond-strength values for slurried steel pipe piles in low organic content silt, together with correction factors to permit estimation of values for other pile and slurry materials. These are empirical results from field tests on piles of 18- to 21-ft (5.5 to 6.4 m) embedded lengths in permafrost.

Under dynamic loading, such as from the effects of explosions, earth-

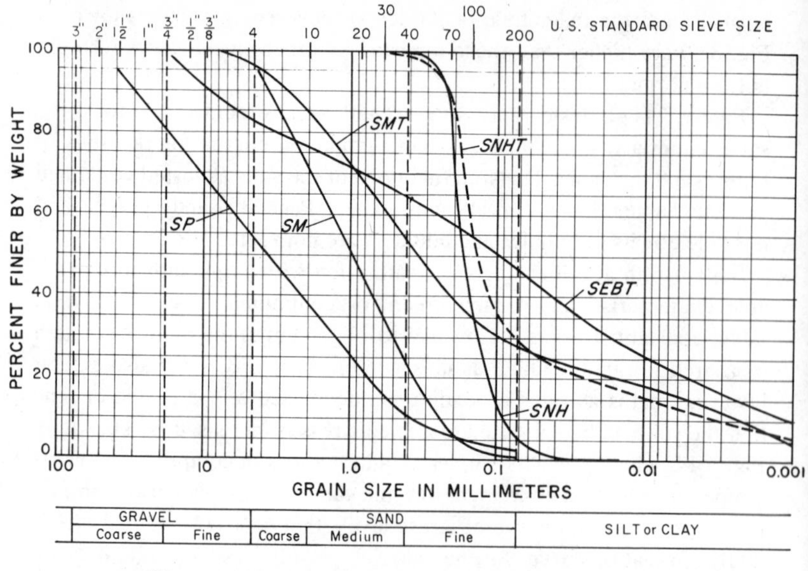

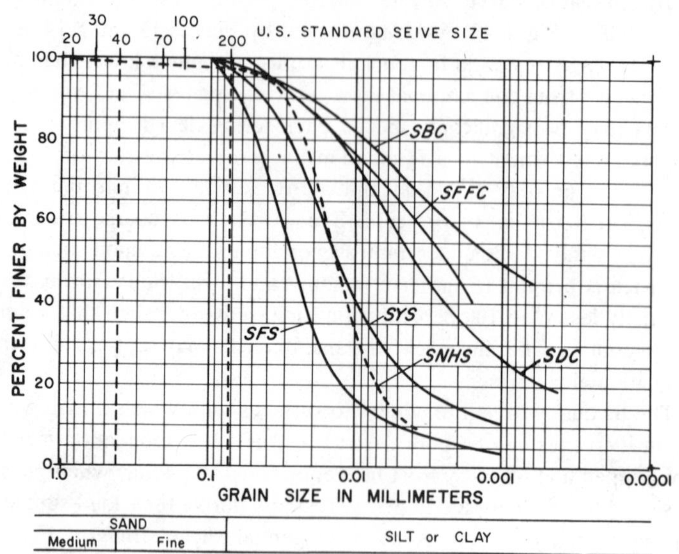

Fig. 8.12. Soil gradations (Kaplar 1954).

<u>Notes for data shown on Figures 8.9, 8.10, 8.11, and 8.13.</u>

Direct shear specimens were 1½ × 1½ × 1½ in. (38 × 38 × 38 mm) and ¾ × 1½ × 1½ in. (19 × 38 × 38 mm).

Compression test specimens were 2¾ in. (70 mm) diameter × 6 in. (152 mm) high, except 6 in. (152 mm) diameter × 12 in. (305 mm) high for Peabody Sandy Gravel.

Tension test specimens were 2¾ in. (70 mm) diameter × 6 in. (152 mm) long, inserted 1½ inches (38 mm) into grips at each end, leaving 3 inches (76 mm) exposed.

Plastic deformation test specimens were 2¾ in. (70 mm) diameter × 6 in. (152 mm) high, loaded in compression.

The clay and peat specimens were tested undisturbed. The remainder of the soil specimens were remolded. All specimens, including ice, were frozen in the laboratory.

Degrees of saturation of soil specimens were between 75 and 100 percent.

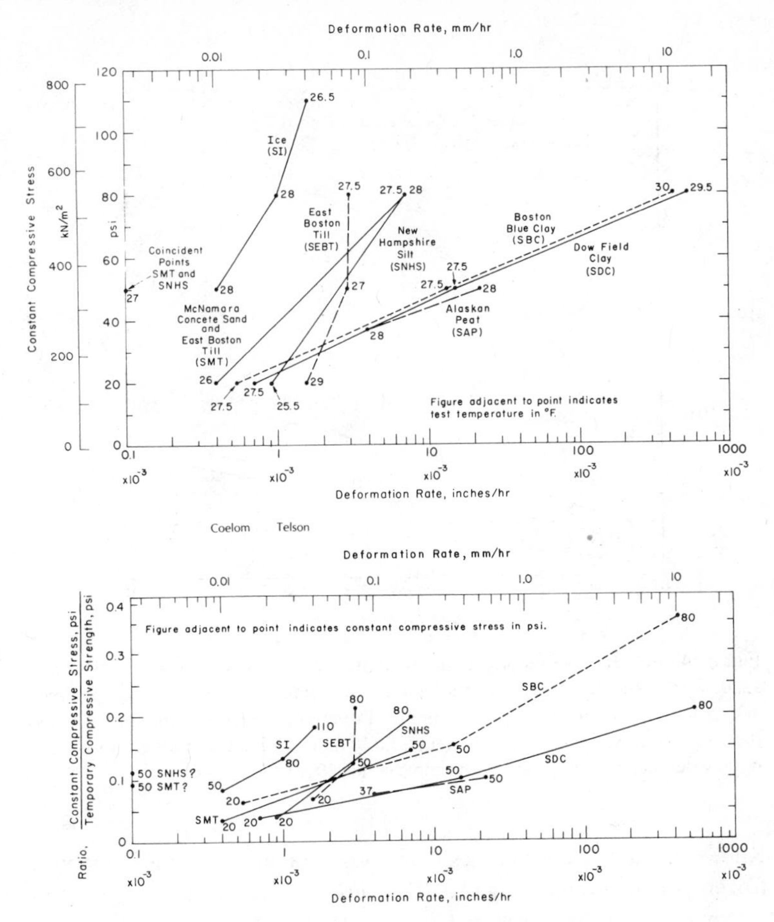

Fig. 8.13. Plastic deformation of frozen soils under constant compressive stress. See Fig. 8.12 for soil gradations (Frost Effects Laboratory 1952).

quake or impact loadings, or relatively low-frequency continuous vibrations, the response of frozen soils and rock differs markedly from that of the unfrozen materials. In general, frozen soils and frozen rock are more brittle, have higher dynamic response moduli, and have less damping capacity than the same materials unfrozen. These properties vary with such factors as temperature, ice volume/soil volume ratio, soil type, load characteristics, and degrees of ice saturation and segregation.

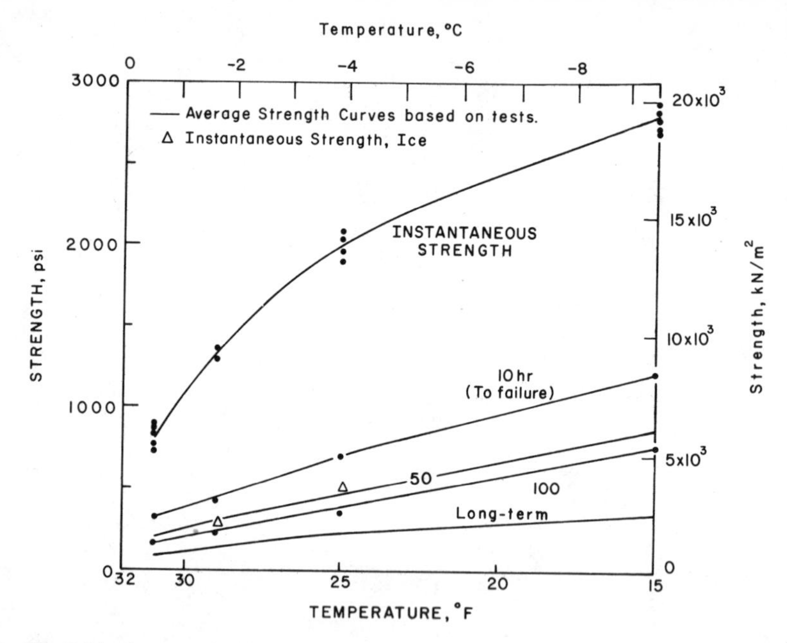

Fig. 8.14. Frozen soil creep tests in unconfined compression on Manchester fine sand. Instantaneous strength is maximum stress determined by loading specimen at constant strain rate of 0.033 per minute. Long-term strength is maximum stress that the frozen soil can withstand indefinitely and exhibit either a zero or continuously decreasing strain rate with time (Sayles 1968).

Knowledge of these properties is important in such activities involving frozen ground as analyzing results of seismic investigations, evaluating the dynamic response of foundations under vibrating machinery, establishing the most economical blasting patterns for excavating soils by means of explosives, or analyzing earthquake effects.

Thermal contraction and expansion

Soils and rocks, in common with other materials, experience significant linear and volumetric changes with changes in temperature. Approximate coefficients of linear thermal expansion are shown in Table 8.1. In the cold regions, intense cooling causes shrinkage and cracking in winter. Shrinkage cracking of flexible pavements and of ice on ponds and lakes is observed in all cold regions, and ground cracking from cold has

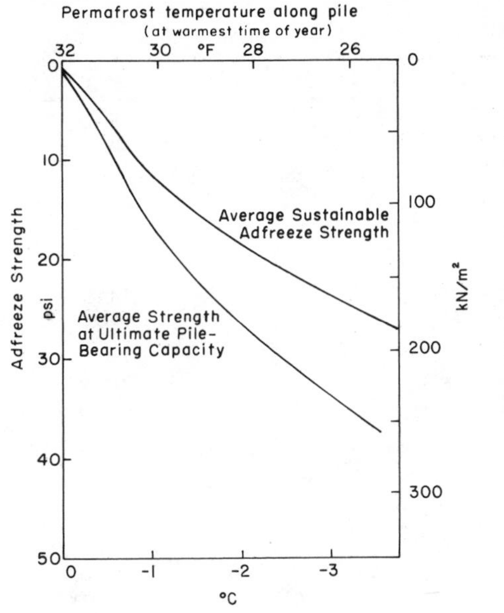

Correction factors for type of pile and slurry backfill (using steel in slurry of low-organic silt as 1.0)

Type of pile	Slurry soil	
	Silt	Sand
Steel	1.0	1.5
Concrete	1.5	1.5
Wood, untreated or lightly creosoted	1.5	1.5
Wood, medium creosoted (no surface film)	1.0	1.5
Wood, coal tar-treated (heavily coated)	0.8	0.8

Notes:

1. Applies only for soil temperatures down to about 25°F (−3. 9°C).
2. Where factor is the same for silt and sand, the surface coating on the pile controls, regardless of type of slurry. In the remaining factors the pile is capable of generating sufficient bond so that the slurry material controls.
3. Gradations typical of soils used for slurry backfill are shown in Figure 8.12 as follows:
   Silt—SFS, Fairbanks silt
   Sand—SM, McNamara concrete sand
4. Pile load tests performed using 10 kips (44.5×10^3N)/day load increment were adjusted to 10 kips (44.5×10^3N)/3 day increment to obtain curves shown.
5. Clays and highly organic soils should be expected to have lower adfreeze bond strengths.

Fig. 8.15. Tangential adfreeze bond strength versus temperature for silt-water-slurried, 8.625-in. (219 mm) OD steel pipe piles in permafrost averaged over 18- to 21-feet (5.49 to 6.40-m) embedded lengths in permafrost (Diagram after Crory 1966, adapted from *Permafrost: Proceedings of an International Conference,* p. 470, with the permission of the National Academy of Sciences, Washington, D.C. Linell, Lobacz, et al. 1980).

been observed in seasonal frost areas as well as the permafrost regions. In the arctic and subarctic, ice wedges typically form at the boundaries of polygons, which are formed in the uppermost layers of the ground. Where accumulation of soil has occurred on the surface, possibly gradually over hundreds or thousands of years, buried polygons and ice wedges, not evidenced on the ground surface, may often be found. Any construction features embedded in or laid on ground subject to seasonal thermal contraction and expansion effects may have stresses imposed upon them. Where items such as power cables or pipes cross contraction cracks, stresses may be sufficient to damage or rupture these members. Structures supported above the surface may also experience such effects if the strains are differential and if they can be transmitted upward

TABLE 8.1.
Approximate coefficients of linear thermal expansion per °C.

	$\times 10^{-6}$
Granite and slate	8
Portland cement concrete	10
Soil, 109 lb/ft^3 (1746 kg/m^3), 23% water content, +20 to −160°C	22
Ice	51
Steel	12
Copper	14–17
Aluminum	18–23
Sulfur	64
Coal tar pitch	160
Asphalt	215
Roofing felt	11–33
Built-up roofing membranes	15–53
Bakelite	22–33
Some other plastics	35–90
Wood (pine), parallel to fiber	5.4
Wood (pine), perpendicular to fiber	34

Note: The coefficient of cubical expansion may be taken as three times the linear coefficient.

Source: Linell, Lobacz, et al. 1980.

through the supporting members. Pipelines and other engineering features exposed to temperature extremes may also, of course, experience severe thermal contraction and expansion stresses directly.

Permeability, drainage, and groundwater

In nonfrost areas the permeability of soils can be an extremely important factor in determining the cost, method of approach, safety, and practicability of many engineering projects and operations. Much expense may be required, for example, to control seepage flow through earth dams and foundations. In civil engineering it is customary to measure permeability by the *coefficient of permeability, k,* defined as the rate of discharge of water under conditions of laminar flow through a unit cross-sectional area of soil under a unit hydraulic gradient. It is the constant of proportionality in Darcy's law, and is usually expressed in centimeters per second:

$$k = \frac{v}{i}$$

where

k = the coefficient of permeability, customarily expressed in centimeters per second.
v = quantity of flow per unit cross-sectional area per unit of time, and
i = hydraulic gradient.

In conventional soil mechanics usage, values of k are expressed at a standard temperature of 20°C (68°F). In the arctic and subarctic, seepage flow normally occurs at lower temperatures, down to 0°C (32°F). The permeability at 20°C (68°F) may be corrected to the permeability at any other temperature by the equation:

$$k = k_{20} \frac{\eta_{20}}{\eta}$$

where

k = permeability at any given temperature,
k_{20} = permeability at 20°C (68°F),
η = viscosity of water at given temperature, and
η_{20} = viscosity of water at 20°C (68°F).

Typical permeability values for unfrozen soils and methods for obtaining the coefficient of permeability are shown in Table 8.2.

When soils freeze, the moisture fraction becomes immobile except for very thin films. Saturated gravels, sands, and silts become, for all practical purposes, impervious as soon as freezing occurs. Even if such soils are significantly undersaturated and retain sufficient connecting voids to have some degree of permeability, they readily become impervious as soon as surface or seepage water reaches them and freezes to plug the remaining voids, provided flow velocities and water temperatures are low. If flow rates or water temperatures are high, however, the ice already in the voids may thaw. Sanger (1968) has analyzed the conditions under which soil strata carrying seepage flow can be intentionally frozen for construction or other purposes. Because clays freeze progressively with lowering of temperature, and moisture can continue to move in unfrozen films behind the freezing front, corresponding progressive changes of permeability with lowering of temperature may be expected in these soils.

In soil strata containing high percentages of dissolved salts, the crystallization temperature and the onset of the effects of ice on permeability may be depressed far below 32°F (0°C). Linell and Kaplar (1959) have reported that in such cases of depressed freezing point, ice segre-

TABLE 8.2.

A summary of soil permeabilities and methods of determination

Coefficient of permeability, k			Relative permeability	Soil type	Method of determination
cm/sec	ft/min	ft/yr			
10 1	20 2	10.5×10^{5} 1.05×10^{5}	High	Clean gravels Coarse sands	
1000×10^{-4} 100×10^{-4} 10×10^{-4}	0.2 0.02 0.002	10,500 1,050 105	Medium	Medium sands Fine sands and sand and gravel mixtures Very fine sand	
1×10^{-4} 0.1×10^{-4} 0.01×10^{-4}	2×10^{-4} 0.2×10^{-4} 0.02×10^{-4}	10.5 1.05 0.105	Low	Silty sands, organic silts Silts, glacial till Silty clay	Computation from grain size Constant head permeameter, reliadle Computation from consolidation test data (reliable) Falling-head permeameter: Reliable / Unreliable / Fairly reliable Field pumping tests, reliable if properly executed
100×10^{-9} 10×10^{-9} 1×10^{-9}	200×10^{-9} 20×10^{-9} 1×10^{-9}	105×10^{-4} 10.5×10^{-4} 1.05×10^{-4}	Practically impervious	"Impervious" soils, e.g., homogeneous clays below zone of weathering	

Source: U.S. Army/U.S. Air Force 1961: Modified after Casagrande and Fadum.

gation and frost heave proceed after onset just as they would if the crystallization temperature were close to 32°F (0°C).

Infiltration of surface water into the ground in nonfrost areas varies with such factors as the type and porosity of the soil cover, temperature conditions, the amount of organic matter in the soil, the vegetative cover, and the degree of moisture saturation of the soil. In permafrost areas, however, the rate of infiltration is usually assumed to be zero in design of drainage and other hydraulic facilities because of the high probability that the surface will be frozen or the active layer will be at 100 percent saturation when maximum rainfall, snow melt, or some combination thereof occurs. The storage of water during the winter months as snow and ice cover and its release in spring and summer by melting profoundly affects moisture availability and drainage conditions in all cold regions.

Subsurface drainage installations are usually ineffective in permafrost areas. Even if placed in the annual thaw zone, the soil around subsurface drains is likely to be frozen and thus impervious to seepage, and the drainage pipes, including any surface discharge points, may be plugged with ice, when functioning of the drains is most needed.

In planning engineering projects, it is of key importance to the engineers to know the position of the groundwater table (including any perched water tables and heads in any aquifer strata) and its variation with the seasons and with changes in the levels of surface water bodies. The position of the water table may limit feasible depths of excavation, for example. If excavations must penetrate below groundwater level, complex earth-pressure and water control measures may be required. The soil gradation and permeability determine which types of groundwater-control procedures can be used in excavations below the water table. Gradations of the foundation soils also determine the required gradations for filter courses needed to prevent soil piping or erosion under seepage conditions in excavations and dams or on slopes.

In permafrost areas, as illustrated in Fig. 8.16, the water table may disappear rapidly during the early part of the freezing season as free water is drawn upward to form segregated ice at the plane of freezing. As further indicated in Fig. 8.16, however, ice segregation and frost heave may continue for some time after disappearance of the water table, supplied by the substantial volume of water still remaining in the voids. During this period, moisture availability and pressure at the freezing plane and the rate of frost penetration are continuously changing.

Soil investigation teams should be alert to the fact that permafrost,

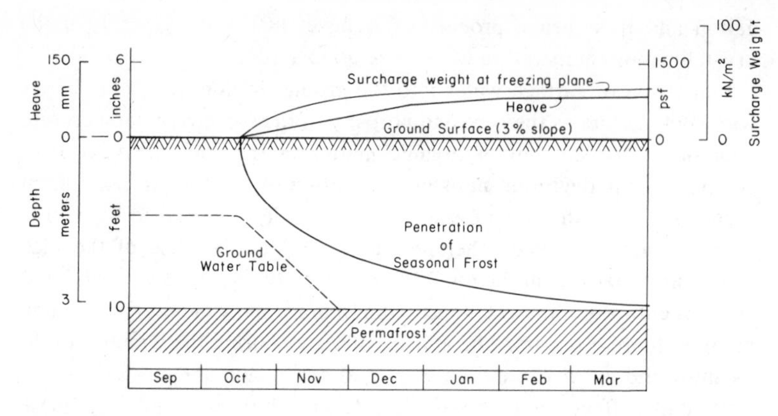

Fig. 8.16. Generalization of changing ground conditions as freezing penetrates the active layer (ACFEL 1954).

acting as an impervious stratum, can cause artesian pressure to develop in groundwater under various geological situations. One of these is illustrated in Fig. 8.17. Water entering the ground at the higher elevations, where permafrost was absent, was trapped under the permafrost at lower elevations. When well no. 3 was drilled at the location shown, hydrostatic head was measured as 8.44 m (27.7 ft) above the ground surface. Flow of water to the surface along the outside of the well casing caused thaw of permafrost and loss of control of the well. Control was regained only through installation of mechanically refrigerated freeze points around the well.

Compaction characteristics

In planning construction work it is important to know the compaction characteristics of soils which may be used in embankments, fills, and backfills. The unit dry weight and relative degree of densification attainable with a given compactive effort affect the physical behavior characteristics of compacted soil. Compaction of soils for construction by rolling, vibration, tamping, or other means is necessary for the following reasons:

To avoid later settlement, especially differential settlement.
To provide required bearing capacity and shearing strength.
To control potentials for moisture gain and frost heave.

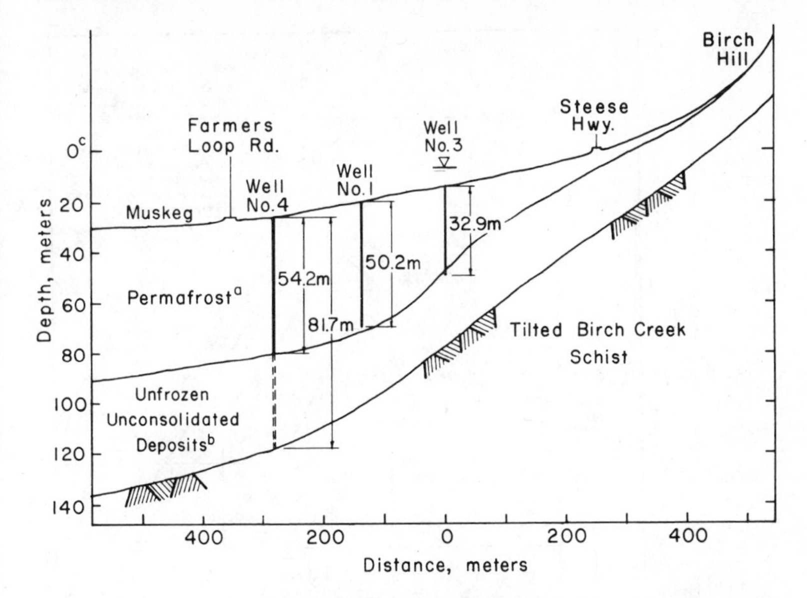

Fig. 8.17. Section from Birch Hill across Alaska Field Station: (a) principally silt with lenses of sand, sandy gravel, and some fractured bedrock; (b) fractured bedrock and silt with lenses of water-deposited sand and sandy gravel; subpermafrost water under hydrostatic pressure. Ordinate depths are measured from Steese Highway datum (Linell 1973a. Adapted from *Permafrost: Second International Conference, North American Contribution,* 1973, with the permission of the National Academy of Sciences, Washington, D.C.).

To achieve determinate permeability and thermal, shrinkage, and swell conditions.

To control earth-pressure loading on buried pipes and other structures.

Standardized laboratory test procedures have been developed for determining the compaction characteristics of soils. A standard compaction effort is applied to the soil in a metal container of standard dimensions. When samples of a given soil are compacted in the laboratory at various moisture contents, an *optimum moisture content* for compaction is found, as illustrated in Fig. 8.18. In the United States the most common laboratory compaction test procedures are those designated as AASHTO T99 and Modified AASHTO T99 (see Appendix A).

In order to achieve required shear strength or other properties, the required compaction may be specified as some percentage of the maxi-

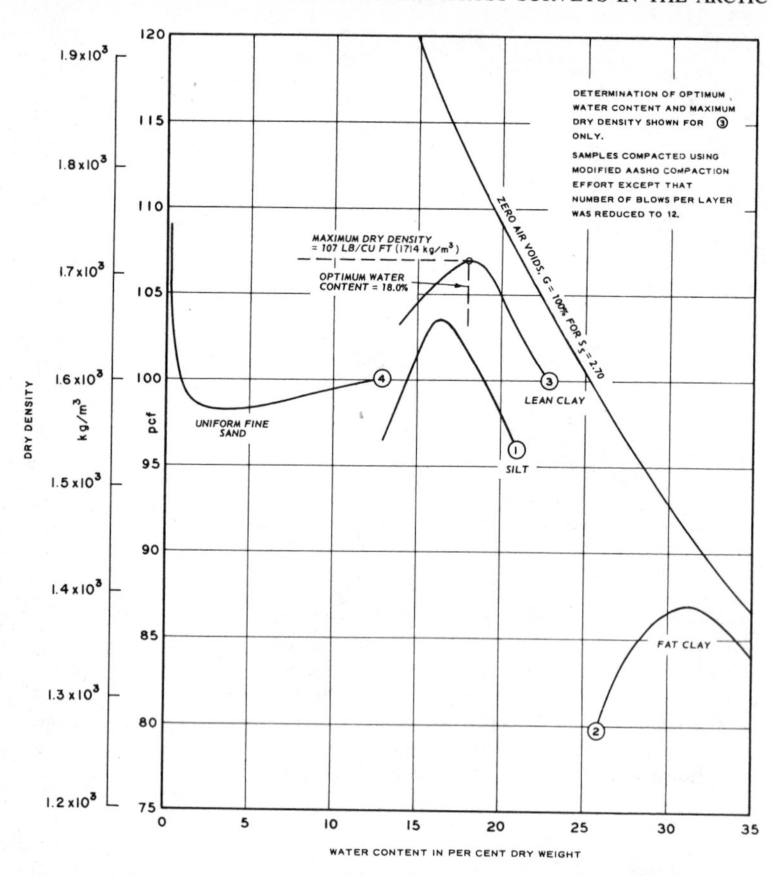

Fig. 8.18. Typical AASHTO compaction test data (U.S. Army/U.S. Air Force, 1961).

mum dry unit weight obtained in the laboratory test, such as 95 or 100 percent. The percentage specified may be varied with the type of soil and with its position in the embankment. In order to achieve the required compaction, the types, combinations, and weights of compaction equipment and numbers of passes can be varied in the field, and the allowable range of moisture contents may be regulated, to produce the specified levels of soil densification. Once a correlation has been established between laboratory and field compaction values for soils of known and uniform characteristics, the requirements may sometimes be simplified to specify only a certain number of coverages of specified types and weights of compaction equipment on specified lift thicknesses of specified moisture content ranges.

Compaction of normally moist, fine-grained soils to normal densities is not feasible when the soils are frozen. Nevertheless, clean, free-draining gravels can be excavated, handled, spread, and compacted to reasonably satisfactory degrees of compaction at below-freezing temperatures even when frozen, if they have in their natural state a very low moisture content or have been dried to such a level before freezing. Occasionally, salt solution is added to unfrozen soil to lower the freezing point of the soil moisture sufficiently to permit placement at below-freezing temperatures.

9. Changes in engineering problems from low to high latitudes

For general purposes the intensity of freezing or thawing conditions is often expressed by the *mean* freezing or thawing index, determined on the basis of mean temperatures. The period of record for which temperatures are averaged to obtain the mean value is usually a minimum of 10 years, although 30 is preferable. The most recent index available should be used. Mean freezing indexes in the Northern Hemisphere are shown in Fig. 9.1.

Design on the basis of mean freezing or thawing conditions does not produce satisfactory results for most engineering projects, however. For example, it would not be acceptable to place water pipes at depths which would be satisfactory for winters with freezing indexes equal to the mean but would permit water in the pipes to freeze in all colder winters. Engineers therefore use the *design freezing (or thawing) index.* For use in designing foundations for average permanent structures, this may be defined as the air freezing (or thawing) index of the coldest winter (or warmest summer) in the latest 30 years of record. For design of permanent pavements, the average air freezing (or thawing) index of the three coldest winters (or warmest summers) in the latest 30 years of record is used, because pavements are usually somewhat more tolerant of heave than structures. If 30 years of air temperature records are not available, equivalent index values may be approximated in various ways, as described by the U.S. Army/U.S. Air Force (1966 *c*). Other statistical bases than those defined above may be used as may be appropriate in specific situations. Figures 9.2 and 9.3 show the general distribution of pavement design air freezing and thawing index values, respectively, in North America (the averages of the three coldest winters, or warmest summers, in 30 years of record).*

Significant seasonal freezing of the ground, with damage to secondary roads, occurs about 1 year in 10 in locations such as northern Texas and Maryland. At somewhat higher latitudes significant freezing is experienced every year, and minimum installation depths for water pipes and

* For engineering project design, to attempt to select design indexes from maps is not satisfactory. Rather, indexes are computed directly from the temperature records of weather observation stations nearest the site. It may be necessary to incorporate allowances for differences in elevation or other factors.

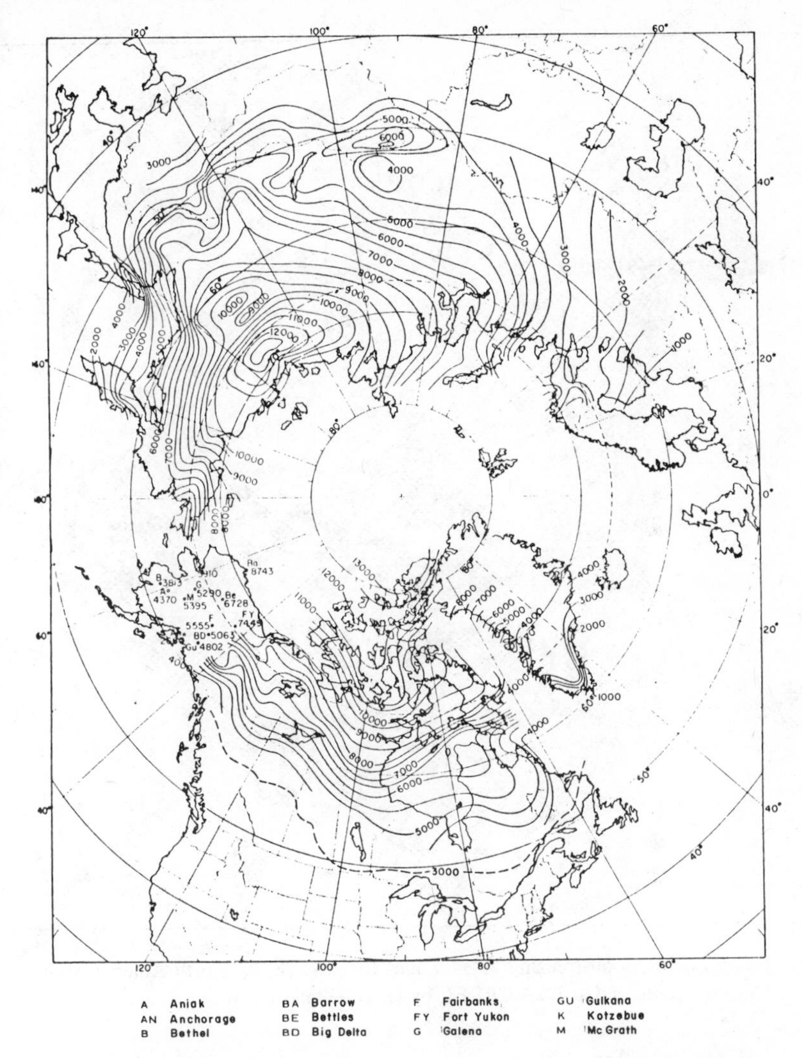

Fig. 9.1. Mean air-freezing indexes (degree days F) Compiled by Wilson (1969) from a variety of sources.

footings have to be specified for frost protection. Substantial measures may be required in pavement construction to control or moderate frost effects. Construction activities are hampered or halted by winter conditions. In still higher latitudes and still colder climates, measures required

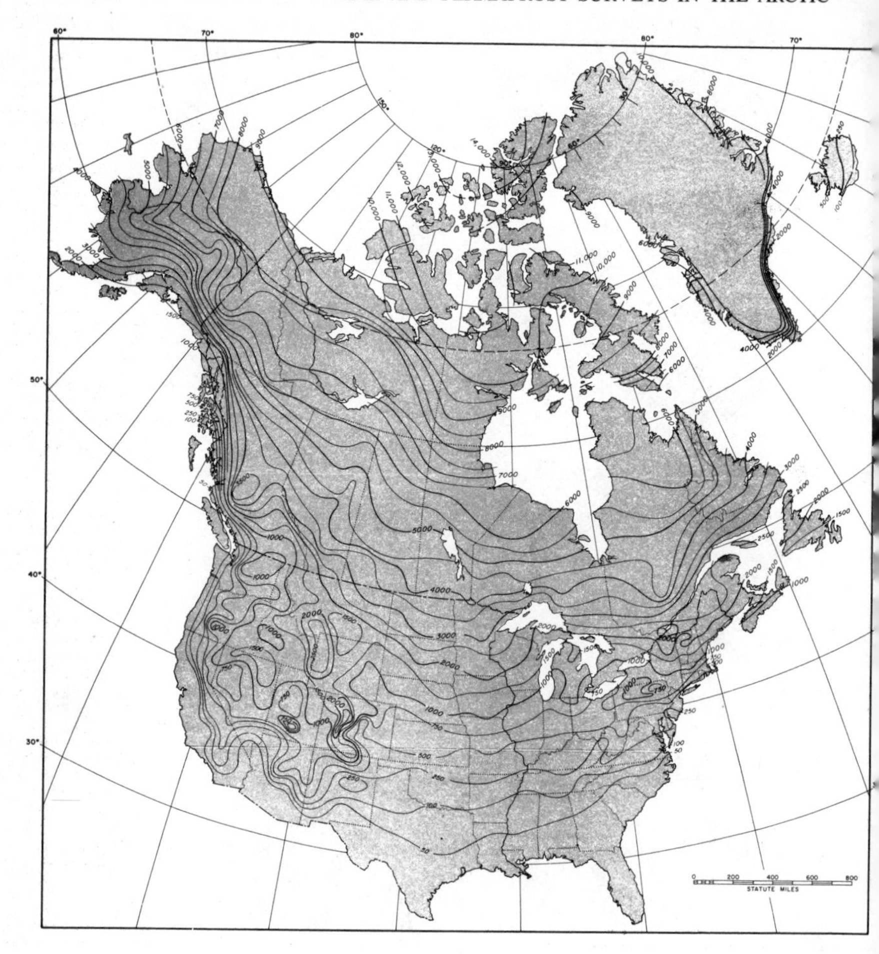

Fig. 9.2. Design air-freezing index values for pavements. North America (Wilson 1969. Prepared at USA CRREL by G. D. Gilman).

to cope with frost effects become increasingly more complex and expensive. At Anchorage, Alaska, for example, municipal water pipes are placed 10 ft (3 m) deep, and heat is added to the water as it enters the system as necessary to prevent freezing in the lines. In still colder locations such as Fairbanks, Alaska, where seasonal freezing may penetrate to depths of 15 to 20 ft (4.5 to 6 m) in nonpermafrost, well-drained gravel, water is recirculated continuously in the mains, with heat added as

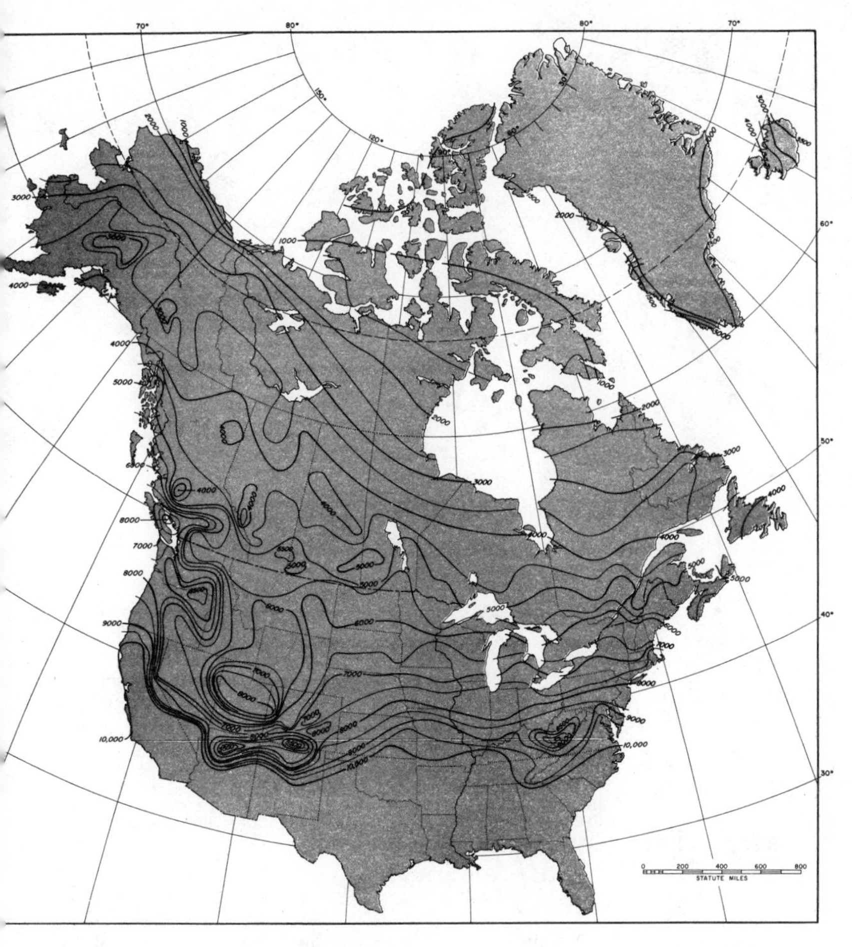

Fig. 9.3. Design air thawing index values for pavements, North America (Wilson, 1969. Prepared at USA CRREL by G. D. Gilman).

necessary. Where conditions favor minimum ground temperatures, permafrost is encountered to depths of as much as 200 ft (60 m) in the Fairbanks area. The progressive change of seasonal frost penetration depth with mean annual air temperature is illustrated by the right-hand curves of Fig. 9.4.

With continuing decrease of mean air temperature as the latitude increases, the depth of seasonal thaw and refreeze over permafrost be-

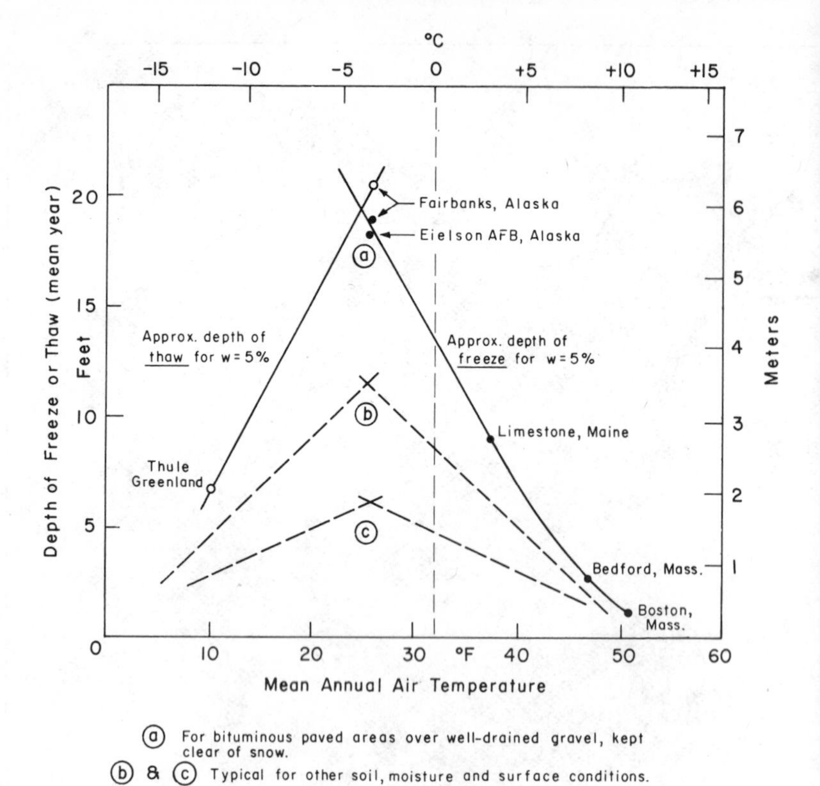

Fig. 9.4. Freeze or thaw penetration versus mean annual temperature (Modified from Linell 1957).

comes progressively thinner, as shown by the left-hand curves of Fig. 9.4. In very cold locations, such as northern Greenland, the annual depth of thaw is very small, and the depth of permafrost is as much as 1500 to 2000 ft (approx. 450 to 600 m). Because of the shallow depth of thaw and the low temperature, thermal stability, and high strength of the permafrost, engineering problems are simpler in many ways in the latter locations than in the marginal permafrost areas farther south.

The most difficult engineering problems in the cold regions occur in the zone of very deep seasonal frost and discontinuous permafrost, approximately under the intersections of the curves on Fig. 9.4. Not only are depths of seasonal freeze and thaw at their maximums here, but permafrost, when encountered, is relatively warm, is easily degraded, has minimum reserve of cold for pile freeze-back, and has minimum strength properties. Further, permafrost may be very sporadic, and small islands or inclusions of permafrost, difficult to detect except with very closely spaced and costly subsurface explorations, may underlie foundations. This situation is compounded by the fact that development is much more intensive in this zone than in the more remote, far northern arctic.

10. Northern agriculture and soil conservation

Northern agriculture

There has always been an interest in the northern limits of agriculture,* particularly in the growth of forages, cereals, and vegetables. The term *North,* has an uncertain meaning—particularly in reference to the Soviet Union. The *Soviet North* generally includes many settled areas of that country, especially within the sparsely forested sectors. The term *Extreme North* (*Krainii Sever*) designates that sector settled by 'northern people' (Slavin 1958). Therefore, when Soviet investigators use the term 'northern agriculture,' they are not necessarily allying their discussions with arctic conditions. 'Arctic' discussions, in reality, often treat the subarctic (taiga or lesotundra) rather than the treeless sectors.

Historically, agricultural development north of the tree line has not shown a great deal of promise (Bliss 1978). Ivanovskii (1963), in evaluating agricultural possibilities in the arctic, correctly stated that in the arctic desert zone (polar desert) there can be no agriculture and within the tundra zone the possibilities are considered negative. Ivanovskii indicated, however, that some possibilities exist within the forest tundra and taiga zones botanical (subarctic).

The lower threshold soil temperature for agricultural crops is generally considered to be 8 to 11°C, and this minimum temperature must be maintained for normal functioning of the plant organism. In addition, if plants are to thrive, this temperature range must be maintained for at least 60 days. Of course, the numerical value changes somewhat with various plants.

Summer soil temperatures in the subarctic commonly exceed 10°C for extended periods of time, but a great variation exists according to soil type, site location, exposure, and related factors. Although grazing animals, such as caribou and reindeer, thrive in most sectors of the arctic, there is no conclusive evidence that significant land areas north of the tree line can be economically tilled. If there is to be further extension of agriculture into northern lands, its development will undoubtedly center primarily on the subarctic and the forest proper.

As a whole, the botanical subarctic region can neither be precisely

* Forestry is not included as a part of agriculture in this discussion.

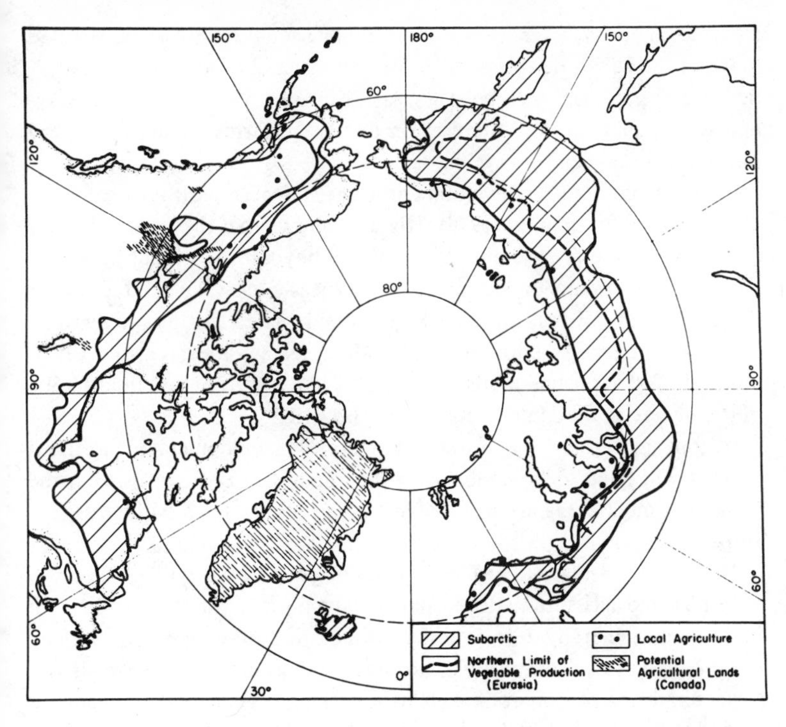

Fig. 10.1. Circumpolar map showing locations of northern agricultural possibilities. Virtually all agronomic success has been confined to the subarctic.

defined nor graphically depicted, although, by defining it as that zone between the forested land having a closed canopy and a polarward extension of conifers, we can draw a general picture. Figure 10.1 shows the botanical subarctic lands in general and notes the locations of agricultural activity. In discussing agricultural possibilities in the permafrost areas of Eurasia, Korol (1955) indicated that in exceptional cases, such as those in the southernmost reaches of the arctic (subarctic), small grains and vegetables may be grown on a small scale. At Naryan-Mar (near the delta of the Pechora River), Igarka (along the Yenisei River just north of the Arctic Circle), and Taimyr, potatoes, vegetables, and cereals have been grown but the yields have been small. Berson et al. (1968) reviewed crop productivity at the Yamal and Khanty Experimental Stations and reported yields to be rather low. Tikhomirov (1962) showed the northern limits of agriculture in Siberia with reference to the tree line, and from his map we can see that agriculture has not penetrated north of the subarctic (pretundra). In Fennoscandia small plots of vegetables and potatoes have been grown as far north as Narvik, Norway, and Kevo, Finland, and there are a few small farms—mainly hay and pasture at Varanger Fjord, Norway.

Potential agricultural development in the northern part of Canada has been outlined by Bentley (1978). His map shows that the possibilities of extending northern agricultural production beyond its present limits are somewhat restricted to the northern Alberta–Great Slave Lake sector (Fig. 10.1). In Alaska, agriculture is limited mainly to the valleys just north of Anchorage, particularly the Matanuska Valley (Gasser 1951). Of the 15,000 acres under cultivation in Alaska in 1976, more than 11,500 of them were in the Matanuska Valley (Bell 1979).

It appears definite that future agricultural development in northern regions will have to be confined to the subarctic. But even within the subarctic, most areas are unsuitable for agriculture for a variety of reasons.

Much of the subarctic has some form of permafrost, which will influence possible agricultural development. The permafrost table tends to be quite deep, generally 1 to 5 m. In muskeg and other types of organic terrain, however, frozen substrate may be within 20 to 30 cm of the surface. Normally, subarctic soils have a thick organic layer, especially in the low positions, which acts as insulation. Once the native vegetation and organic matter have been removed, however, the thermal regime of the soil changes, resulting in increased heat conduction and subsequent lowering of the permafrost table (Fig. 10.2). Uneven thawing of the underground ice structures causes collapse and erosion of the earthy material. Locations having large quantities of ground ice present in the form of ice wedges and massive ground ice structures are most susceptible to subsidence and collapse.

Conservationists frequently and correctly speak of the need to protect the vegetative cover in arctic regions because it prevents deterioration and subsequent erosion of the frozen soil. If the vegetative cover is destroyed there is a tendency for accelerated erosion to set in. Figure 10.3 shows such a situation within the taiga of Siberia which was stripped of the vegetative cover. Differential thaw of the ice-wedges resulted in the development of a network of troughs surrounding polygonal remnants. In surveying subarctic soils for possible agricultural development, it is therefore important to study the distribution of ground ice as well as the soil cover itself.

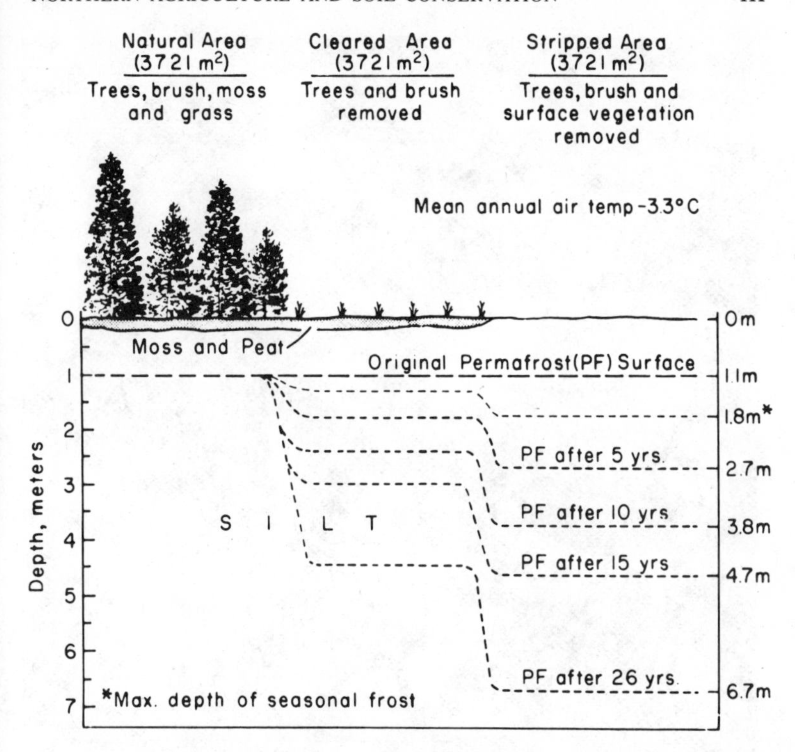

Fig. 10.2. Permafrost degradation under different surface treatments over a 26-year period. Mean annual permafrost temperature ranges from about −0.5°C at a depth of 10 m under natural forested areas to about −0.2°C where permafrost is degrading. (Linell 1973b. Reproduced from *Permafrost: Second International Conference,* p. 691, with the permission of the National Academy of Sciences, Washington, D.C.).

Natural depressions sometimes form in the landscape from the melting of the underground ice (Czudek and Demek 1970). These depressions, referred to as *thermokarst, thermokarst depressions,* or *alasses,* are sometimes used to the advantage of agriculture, especially in Siberia. Most of the subarctic is considered moisture-deficient during the growing season, but once the depressions form, the soils tend to have a more favorable moisture regime. Accordingly, some Yakutian farmers tend to settle near alasses where there is a more reliable source of soil moisture for crop growth (Fig. 10.4).

Fig. 10.3. View of a Siberian roadway from which vegetation was removed and ice wedges consequently melted.

In most of the subarctic drought will seriously impede potential agriculture. Much of the subarctic receives only about 30 cm of precipitation per year. According to Sanderson (1948), drought is common in northwestern Canada, a condition also found throughout most of the Soviet subarctic—particularly in central Siberia.

On sandy deposits, such as deltaic and outwash, conditions may be well-drained to xeric, with well-developed Podzol soil being present. Such soils, however, generally do not have an adequate moisture-holding capacity for reliable crop growth. Further, when the forest canopy is broken on sandy deposits and the organic soil horizons are destroyed the wind-erosion potential becomes critical.

Fig. 10.4. View of a thermokarst basin (alass) in Siberia. The ground ice melted, causing the land to subside. The basin has a favorable moisture regime and fosters agricultural development of the land.

Saline-alkali soils are present in the subarctic as well as the arctic. Two large sectors, one in northern Alberta and one along the middle course of the Lena River, commonly have enough salt present to restrict potential agricultural production. Many of the subarctic saline-alkali soils of Siberia are underlain by permafrost, but this is not always the case in western Canada (Tedrow 1970).

Muskeg is a term describing wet, peaty terrain, including such conditions as fens, bogs, swamps, and marshes. Distribution of muskeg is mainly subarctic, but it is also present in the boreal forest as well as the arctic.

As a result of the insulating qualities of organic material, the southern extremities of permafrost are usually associated with muskeg. There are, however, extensive areas of muskeg without permafrost. The statement of Linnaeus, 'Never can the priest so describe hell, because it is no worse,' portrays field conditions in the vicinity of muskeg. There are many varieties of organic material which make up the peaty soil in muskeg areas, but basically it is derived from plant residues in wet depressions. Also along the margins of shallow bodies of water the infilling of living and dead plant material forms the peaty substrate. Though there are a number of origins of muskeg, they all have one property in common—wet, organic terrain (Radforth and Brawner 1977).

Pihlainen (1965) made construction recommendations for muskeg sites and emphasized that muskeg should be avoided if possible; if not possible, the peat should be removed and replaced by fill to provide a solid foundation, after which specially designed construction may proceed.

Soil conservation

Only within the past decade or so has the problem of soil conservation, including erosion control, in the arctic been recognized as important. Prior to the advent of major construction in the arctic, there was little need to consider soil conservation measures because, over the centuries, nearly all arctic lands had remained largely in their natural state. In the aftermath of increased activity in the arctic, however, particularly the building of roadways, pipelines, and airfields and other forms of construction, the picture has changed. Conservation measures which were largely developed for temperate and warm climates are not always effective in arctic lands. The problem of erosion control in arctic lands, for example, has a number of unique dimensions—particularly the presence of permafrost, low vegetative recovery rate, low soil temperatures, and a short growing season, among others.

Cutslopes and other man-made, raw, mineral soil sloping sites pose special erosion hazards. Considerable difficulty is generally encountered in revegetating such newly exposed sites. The problem of erosion control is especially acute where there is fine-textured material with high quantities of ground ice present. In addition to the erosion hazard *per se*, there is also a potential problem created by the downstream deposits of the eroded mineral material.

In classifying arctic soils and landscapes for erosion susceptibility, some of the factors to be considered are:

(i) Soil texture (including stoniness). Arctic soils are predominantly silty in character, but textures range from fine clays to sands. The degree of stoniness ranges from stone-free to block fields (felsenmere).
(ii) Genetic soil variety. Well-drained conditions should be distinguished from those with poorly drained bog or shallow conditions.
(iii) Ground ice. This is one of the more critical aspects to be considered in the delineation of sites. With large percentages of the soil matrix consisting of ice, it is important to indicate some approximation of the ice content together with its distribution.
(iv) Slope of the land.
(v) Vegetative cover.

One of the major concerns of those engaged in erosion control in the arctic should be maintaining the thermal balance within the soil. In cases where the vegetation is destroyed the heat regime within the soil is altered, which commonly results in a progressive thaw-erosion cycle taking place. It is difficult to establish effective erosion control measures when there are gley soils on sloping lands of the arctic because of the instability of the soil. During construction of the Alaska Pipeline, approximately half of the approximately 800 miles (1280 km) of line crossed land with some form of permafrost; accordingly, special precautions had to be taken in major erosion-susceptible sectors. In order to avoid thawing the permafrost in critical areas, about 382 miles (611 km) of the pipeline were built above ground. Lachenbruch (1970) calculated that pumping hot oil (~80°C) through buried pipe in permafrost would thaw a cylindrical region 20 to 30 ft (7 to 10 m) in diameter in a few years and, at the end of the second decade, thawing depths would be as much as 50 ft (17 m) near the southern limit of permafrost.

On flat terrain there are various quantities of organic matter accumulation in the soil, some of which may be at least several meters thick. These flat areas are sometimes used as roadways by tracked vehicles. During winter months when the soil is completely frozen and snow covered, there are few immediate erosion hazards or stability problems. During summer months, however, the organic layer becomes macerated from vehicular traffic (Fig. 10.5). As devastated as such roadways appear, the erosion problem itself is not great where the terrain is flat (slopes ~ 1 to 2 percent), but the problem becomes more critical on steeper slopes. Gersper and Challinor (1975) studied soil disturbance on flat, organic-like terrain in northern Alaska and found the soil morphol-

Fig. 10.5. View of a "roadway" in the tundra near Barrow, Alaska. The low, water-covered positions form from the deterioration of the ground ice (courtesy P.L. Gersper).

ogy only slightly altered, the prime differences being the alteration or destruction of the organic surface soil and vegetative cover.

Lawson et al. (1978) analyzed the disturbance and subsequent recovery of soil, vegetation, and permafrost in northern Alaska. At the Fish Creek drill site of (U.S.) Naval Petroleum Reserve No. 4, in 1949, considerable local disturbance of the terrain followed normal drilling operations. After nearly 28 years the terrain was reassessed. The most lasting disturbance of the landscape was from bulldozing surface materials, roads and trails; subsidence resulting from the thaw of permafrost and thermal erosion. Marks made by heavy equipment remained imprinted on the tundra, but the vegetation had closed on most of the mesic sites and the wet sites had revegetated well; xeric sites, as well as those spots of accelerated erosion, such as roadbanks, were slow to recover, however. Overall soil erosion was insignificant. Literature on revegetation of the northern areas was compiled by Johnson and Van Cleve (1976). Their report covers the uniqueness of vegetative restoration research in cold climates.

In arctic soil surveys it is desirable to think of muskeg as a place which should exist in its original condition, disturbing the soil as little as possible. Apart from being unsuitable terrain for vehicular traffic and construction, most muskeg usually exists in some form of low-lying, sumplike positions, which are potentially vulnerable to contamination from oil spill and waste disposal.

The high arctic

At high latitudes the landscape is more barren and summers are shorter, cooler, and drier than in the main tundra belt. During the early summer thaw, much of the terrain is exceedingly wet for a period of weeks, and there is some susceptibility to soil flow. But by midsummer, some soils become quite dry and salt-encrusted. Whereas so much of the rolling landscape of the main arctic belt (tundra) is highly susceptible to erosion, many soils in the high arctic tend to be much drier and appear to be less erosion prone.

Babb and Bliss (1974) approached the conservation problem by recognizing various needs within the high arctic and, accordingly, divided the landscapes into (i) polar deserts, (ii) subpolar deserts, (iii) diverse terrain, and (iv) large meadows. Barnett, Edlund, and Hodgson (1975), in their approach to classifying arctic terrain, stated that to assess the conservation problem properly the following criteria should be considered: (i) surface materials—ice content, texture, and engineering properties; (ii) topography and landforms; (iii) geomorphic processes; (iv) drainage; (v) vegetation; (vi) summer temperature and soil moisture; and (vii) wildlife.

A number of natural erosive forces also operate in the arctic. Solifluction and mass-wasting occur especially on sloping, silty Tundra soils with a large content of ground ice. On steeper slopes, the erosion magnitude is of a high order. Also the higher alpine and arctic lands, as exemplified by conditions in northern Fennoscandia, commonly receive sufficient precipitation to induce a high degree of erosion. Some areas may receive as much as 300 to 400 cm of precipitation annually.

Natural erosive processes may sometimes cause ice wedges to melt resulting in a collapse of land.

Erosion of stream banks is quite active in certain arctic areas. Braided streams undergo considerable annual change from the cutting and filling processes; but these processes are natural.

Aeolian deposits are common in many sectors of the arctic. They occur as blowouts, dunes, and related geomorphic forms. The material usually ranges in size from fine sand to silt and has very poor binding qualities. Under such conditions plants colonize the substrate very slowly, resulting in semipermanent barren spots.

11. Thermal stability of permafrost

The environmental impacts of any disturbances to permafrost terrain, including the effects of survey and construction activities, must be carefully anticipated and controlled. Disturbance of vegetation, drainage, water quality or temperature, together with soil erosion, siltation of streams, and spills of oil or other substances may have profound adverse effects upon the terrain, permafrost, animal and aquatic life, and constructed works. The integrity and permanence of constructed facilities and the utility of the land for future generations may be endangered. Survey, construction, and operating procedures must be planned to avoid man-made contributions to adverse changes. Disruption of the thermal stability of permafrost is a major contributing factor in adverse terrain effects in permafrost areas.

Causes of thermal changes

Within depths significant to engineering projects, permafrost under undisturbed terrain most commonly exists in a reasonable state of thermal equilibrium with the climatic environment, with specific ground temperatures varying with the effects of covering materials. A forest fire, which destroys the surface vegetative materials, can, nevertheless, cause an abrupt change in the thermal regime at the soil surface, and a great many years may pass before the original ground-temperature conditions are reestablished. Again, meandering or shifting of a stream channel can cause substantial changes in subsurface soil temperature and in permafrost conditions, as illustrated in Fig. 11.1.

Man-made changes in the thermal regime and consequent changes in permafrost stability may be initiated by actions such as the following:

Damage to surface cover sensitive to mechanical disturbance, such as by vehicular movements on the surface.
Clearing of trees or brush.
Stripping and removal of surface vegetative mat.
Excavation into and exposure of underlying soil or rock.
Construction of an embankment, road, or runway.
Performance of seismic investigations with explosives in summer.

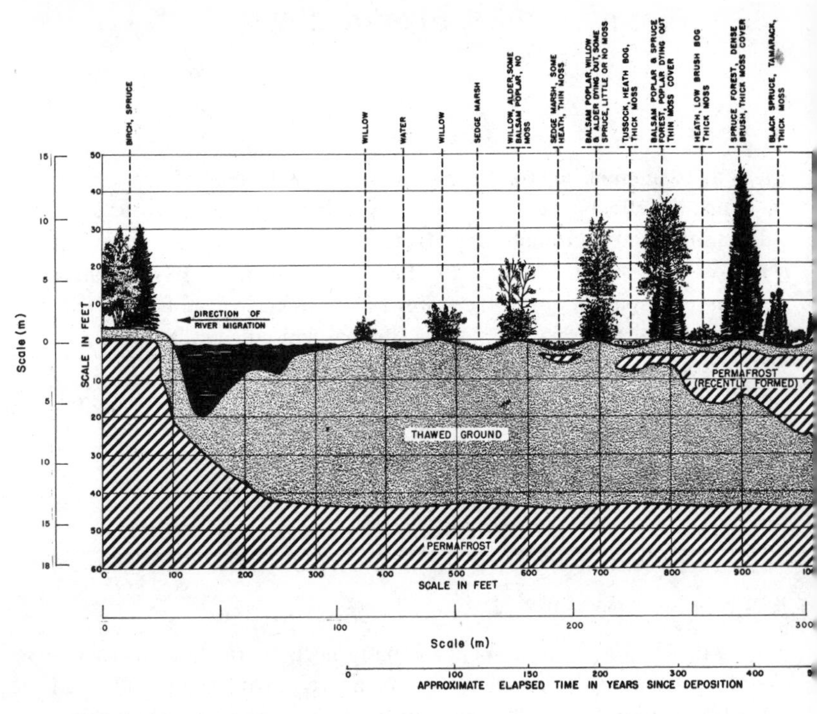

Fig. 11.1. Effect of a stream meander on permafrost. An idealized cross section of a point bar illustrating the relation between permafrost aggradation and time. The area depicted is in a zone where the mean annual temperature is approximately −6.7°C(20°F). Permafrost in the point bar has been destroyed to a depth of 40 ft (12 m) by the thawing action of the stream. The cover types established on each of the point-bar ridges and intervening swales represent the natural process of plant succession typical of this terrain type. With the passing of several centuries, the vegetation cover has become denser, and the associated layer of insulating peat has become thicker. Microclimatic and local environmental characteristics have simultaneously altered sufficiently to make conditions favorable for the local redevelopment of permafrost. In a period of approximately 500 years the new permafrost layer has grown to a thickness of 25 ft. (7.6 m) Under the present climatic conditions, permafrost will continue to penetrate deeper until the upper surface of the second permanent permafrost layer is reached in, perhaps, 1000 years (U.S. Army 1963).

Change in capacity of surface to reflect or absorb solar radiation.
Oil spills.
Change of snow depth or compaction.
Construction of a heated structure.
Construction of a structure which shades the ground.
Discharge of warm water into the ground.
Placement of utility lines or pipelines in the ground.

Whether or not settlement, developing surface unevenness, slumping, wet ground-surface conditions, or other effects which may accompany permafrost degradation will significantly affect such activities as growing of crops or the integrity and usability of pavements, structures, or utilities depends upon the amount of excess ice present within the depth of thaw and its distribution. Serious damage to the terrain itself is limited to soils containing excess ice. Damage to construction features from degradation occurs primarily in soils containing excess ice, but it is also possible in soils in which ice is *not* excess, as when foundation-supporting capacity or frost-heave resistance depends upon the adfreeze strength in permafrost. When degradation continues year after year and the ground contains massive ice, catastrophic subsidence may occur, with settlements often measured in feet, together with progressive erosion, slumping, and destruction of terrain. During thaw, whether seasonal or continuing, settlements tend to vary from point to point, not only because of variations in ice content or consolidation characteristics of the soil, but also because of variations in the soil thermal properties, causing differences in the rate of thaw.

Field personnel should be alert to perceive and record evidences in the field of changes that have occurred in the past, are currently occurring, or may potentially occur in the future.

Effects of modification or removal of vegetative cover

One of the most common causes of permafrost degradation is damage to or removal of surface vegetation, as by vehicular travel or construction operations, even without direct disturbance to the soil itself. The results of an experiment at the U.S. Army Corps of Engineers' Alaska Field Station, Fairbanks, Alaska, illustrated in Fig. 10.2, showed that merely removing a dense cover of 9- to 15-m (approx. 30 to 50 ft) high white and black spruce trees and high brush from somewhat organic silty ground with about a 4 percent slope, while retaining a thick mantle of moss, grasses, and low shrubs, was sufficient to produce about 3.5 m (11.5 ft) of permafrost degradation after 26 years. During the same period permafrost remained stable in an adjacent, tree-covered control section.

Other plots in the Fairbanks vicinity, cleared and maintained as cropland or pasture for a number of years, have degraded to as much as about 7 m (23 ft) in 24 years (Kallio and Rieger 1969). Péwé (1954) has described experience with areas underlain by large ice masses at Fairbanks, Alaska, which were cleared and cultivated in the first half of this century and underwent such severe degradation effects that they became difficult or impossible to farm. Mounds generally 20 to 100 ft (6 to 30 m) in diameter and pits up to 20 ft (6 m) deep were formed. Péwé estimated that permafrost containing large ice masses underlies 27 percent of the Fairbanks area. That permafrost degradation involving excess ice is occurring is often visually apparent in recently cleared areas because of the excessively wet condition of the ground in dry, late summer weather and by the developing unevenness of the surface.

These experiences are representative of a substantial variety of soil, drainage, vegetative cover, and exposure conditions in the Fairbanks area. Nevertheless, the results might vary in the same area with differing drainage conditions, kinds of vegetation and growth characteristics, types of soil and organic mat, conditions of wind, snowfall, and elevation, and degrees of shading or exposure to daily sunshine. In fact, the permafrost in an area close to the plots that yielded the data shown in Fig. 10.2 that was naturally lacking in the dense tree growth that was original cover in the test plots but was level, wet and swampy showed no degradation during the same time period shown in the figure.

Different results may also be experienced in other areas with colder air and permafrost temperatures, differing degrees of cloudiness or windiness, differing amounts of snow cover, or other differences. For example, although the Fairbanks experience has shown that a cover of low vegetation is not sufficient under many conditions to control permafrost degradation in that area, it can be sufficient in a colder location such as Inuvik, N.W.T., Canada. Again, at Schefferville, Quebec, with only slightly colder mean annual air temperature than Fairbanks, 23°F (−5°C) versus 26°F (−3.3°C), permafrost is widespread in the uplands but is common in the lowlands (Granberg 1973), whereas, in the Fairbanks area, permafrost occurs in the lowland areas and is absent at some higher locations. At Schefferville the very heavy winter snowfall is redistributed off the uplands by wind, leaving only shallow snow cover over large areas of terrain, but snow cover is more uniform and deeper in lowland areas, acting as an insulator. At Fairbanks, the coldest winter temperatures occur during periods of calm, and temperature inversions cause the lowest temperatures to occur at the lower elevations. On granular soils in the high arctic, vegetation and organic cover may be so slight as to have no practical effect on permafrost conditions.

Effects of travel on the surface

In some areas where conditions are especially fragile, such as where massive ice is present under only a thin cover of active layer soil and where the soil is easily erodible, even a single pass of a surface vehicle may be enough to initiate progressive permafrost degradation, slumping, and erosion. Even if slumping and erosion do not develop, the tracks made in tundra vegetation by such passage may remain clearly visible for many years, sometimes with water standing in the tracks in summer. Although vegetation killed by such passage perhaps retains some small insulation value, its function for heat exchange through evapotranspiration is destroyed. At the same time, absorption of solar radiation by the standing water adds heat input. If excess ice is present, the resulting degradation and settlement allow more ponding of water, and the situation develops progressively. This is one way in which *thermokarst* features may develop (see Chapter 16).

Effects of water

Both surface water and moving subsurface water are important sources of heat for permafrost degradation. Water temperatures as high as 70°F (21°C) have been measured in the upper 2 ft (0.6 m) of water in shallow lakes as far north as Thule, Greenland, in the summertime under the influence of long daily hours of nearly continuous heat input from solar radiation. Water which seeps into soil or is carried in drainage ways at such temperatures can cause very rapid thawing of permafrost. Discharge onto the ground of waste water or steam condensate from buildings can also be very serious. At a communications building at Glennallen, Alaska, a very slow drip of condensate from a joint in an insulated, overhead steam line, near its entrance into the building, caused degradation of permafrost to a depth of about 18 ft adjacent to the foundation before the drip was corrected.

Water in lakes and rivers which are sufficiently deep may not freeze to the bottom in winter. This prevents freezing temperatures from reaching the underlying ground from above. Consequently permafrost may be expected to be absent immediately at the bottom of such water bodies, although frozen ground may be found below the unfrozen strata as a remainder following stream meander, as illustrated in Fig. 11.1, or as the result of the penetration of freezing temperatures from the edges of the water body. The unfrozen lake or river water and the water in the unfrozen soil can serve as sources of water supply for arctic communities even in the coldest periods of the winter.

Effects of solar radiation

Solar radiation is a very important contributing factor in the input of heat in the arctic and subarctic in the summer. The surface of a pavement is highly absorbent of solar radiation and, as shown by Table 4.1, has a high n-factor in summer. Thus, maximum depths of summer thaw penetration occur under pavements. When pavements are constructed over permafrost, thaw of permafrost may occur unless sufficient thicknesses of gravel or other materials are used beneath the surfacing so that the summer thaw will not reach the permafrost. In the warmer permafrost regions, the thawing index for pavements is so large that it is usually impractical to provide a sufficient thickness of granular, non-frost-susceptible base to protect the permafrost from thaw, the required combined thickness being as much as 15 to 20 ft for bituminous pavements. Thus, degradation of permafrost normally occurs under pavements in these regions, continuing progressively for an indefinite number of years unless special design measures, described in following paragraphs, are used.

New roads over thaw-unstable permafrost are customarily surfaced with gravel because the differential settlements of the surface that occur from thawing of underlying ice can be easily corrected with a blade grader as they develop. Later addition of a bituminous surface, after the gravel-surfaced road has appeared to stabilize, often initiates a new cycle of degradation and settlement, and a newly paved section of road can sometimes become barely passable within only a few weeks in the summer. Blade grading, however, is then no longer possible. In very cold locations, such as Thule, Greenland, about 6 to 7 ft (1.83 to 2.14 m) of combined thickness of pavement and very low moisture content base (average moisture content about 3.5 to 4 percent) is required for full protection of the permafrost under bituminous surfacing. This is still a substantial thickness of material, though tolerable for some features,

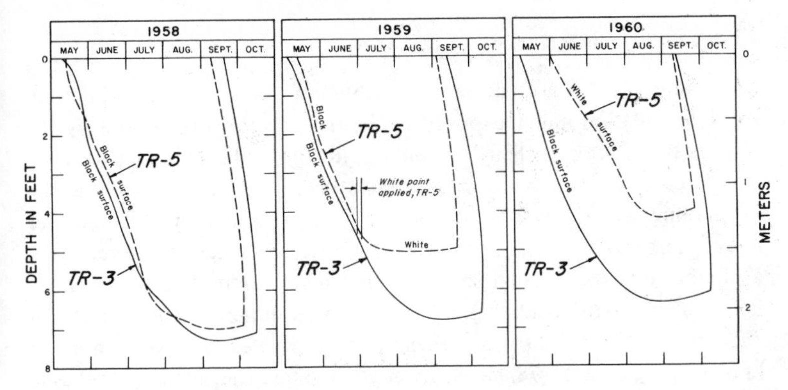

Fig. 11.2. Reduction of thaw penetration by application of white paint on runway, Thule, Greenland, as measured by thermocouples (adapted from Fulwider and Aitken 1962).

such as runways of major airfields. Because the active layer is free of massive permafrost ice, its thickness is often included as part of the required thickness, provided it does not introduce unacceptable seasonal heave characteristics.

Effects of solar reflective surfaces

The U.S. Army Corps of Engineers, in experiments in the late 1940s and early 1950s at both Fairbanks, Alaska, and Thule, Greenland, found that the depth of thaw penetration and the required thickness of cover for protection of underlying permafrost could be substantially reduced by painting the pavement a highly reflective white color (Fulwider and Aitken 1962). As shown in Fig. 11.2, this reduced the depth of thaw penetration at Thule by up to a third. Based on these experiments, the U.S. Air Force later painted the runway and other critical pavements at the Thule airfield with a white traffic paint which contained ground pumice to provide skid resistance. The results were fully successful. Progressive reduction of albedo in following years was reasonably slow. In later experiments at the Corps of Engineers' field station at Fairbanks, Berg and Aitken (1973) verified that painting pavements white can be effective in controlling and limiting permafrost degradation. In their experiments the permafrost table under the white-colored pavement remained almost stationary at not over a 2.5-m (8.2 ft) depth for 5 years of observation, after 1 year of initial adjustment.

Effects of thermal insulation

Thermal insulation provides an alternative means for controlling permafrost degradation under exposed surfaces. Experiments started by the Corps of Engineers at their Fairbanks field station in 1947, using cellular glass, cell concrete, and fiber batt insulating materials, showed that the insulation was initially effective in controlling degradation but that after a few years the normal rate of degradation was resumed. More recently, a number of experimental and operation installations have been made using foam plastic insulating materials. These materials show excellent short-term effectiveness, but their long-term effectiveness is not yet known. Berg et al. (1978) have recently compared the calculated relative effects of gravel, bituminous pavement, white-painted pavement surfaces, and in-ground insulation on maximum seasonal thaw penetration below the surface of the Livengood to Prudhoe Bay Road, Alaska.

The surface color modification technique controls thaw penetration and degradation by reducing the surface thawing index and the mean annual surface temperature, but its effectiveness is limited primarily to minimizing radiant heat input. The thermal insulation technique does not alter the solar heating effect. Instead, it interposes a thermal barrier to limit the penetration of the incoming heat into the ground in the summer; however, it also limits the penetration of freezing temperatures in the winter. The net effect may be a long-range warming of underlying permafrost, disregarding edge effects. Of course white-painted pavements present the problem of maintaining their whiteness under traffic, considered to be more of a problem on highways than on airfields. Still, the principle of using reflective surfaces to minimize solar radiation heat input is not limited to use on pavements. Wechsler and Glaser (1966) and Kritz and Wechsler (1967) have summarized and analyzed available information on the effect of surface characteristics on thermal regime.

Effects of buildings and other facilities

Figure 11.3 shows diagrammatically the effect of size on both total depth of thaw and rate of thaw under a uniformly heated structure placed directly on a frozen material of uniform properties. Thawing proceeds most rapidly near the center of the structure. The resulting thaw front is bulb-shaped, and the settlement surface from melting of uniformly distributed ground ice is dish-shaped. The larger the structure, the larger the potential ultimate depth of thaw. At the start of thaw, however, the

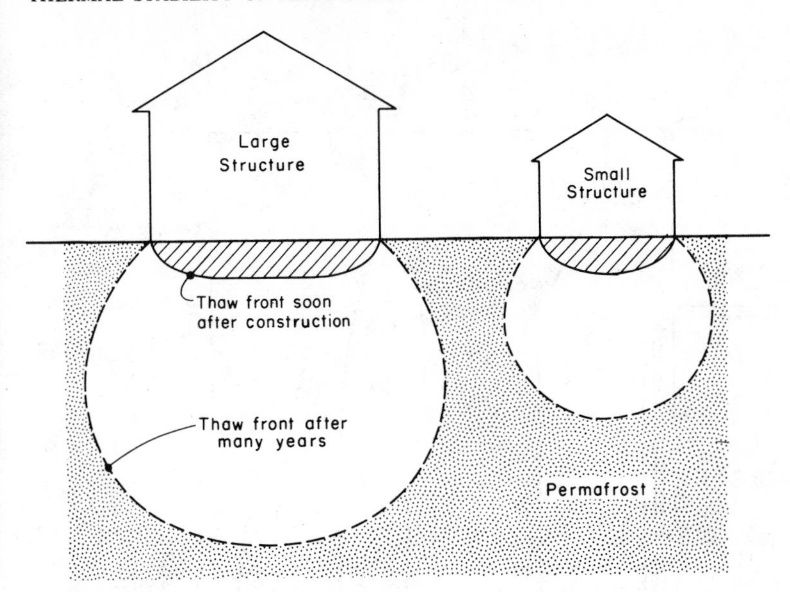

Fig. 11.3. Effect of size of heated structure on depth and rate of thaw (Linell, Lobacz, et al. 1980).

rate of thaw advance under the center of the structure is not a function of structure size. Actually, settlements can be quite variable over the area of the foundation because of variations in ice distribution in the soil. Settlements under basement floor slabs may not be detected because of bridging of the floor slabs.

If a building or other constructed feature does not transmit heat into the ground, but rather shades the ground from solar radiation in summer and allows free contact of cold air with the underlying ground surface in winter, or if it is refrigerated to below the normal ground temperature, the permafrost temperature at that location may be lowered and the permafrost may *aggrade*. On the other hand, a building or other structure which emits heat to the ground can cause progressive degradation of the permafrost over a period of years, as illustrated in Fig. 11.4(a). If the permafrost does not contain excess ice and is thaw-stable, the degradation will not cause any settlement except that due to the compressibility of the unfrozen material, although water problems may occur in basements or other below-ground construction because of the sump effect of the resulting thaw pocket. If the permafrost contains excess ice,

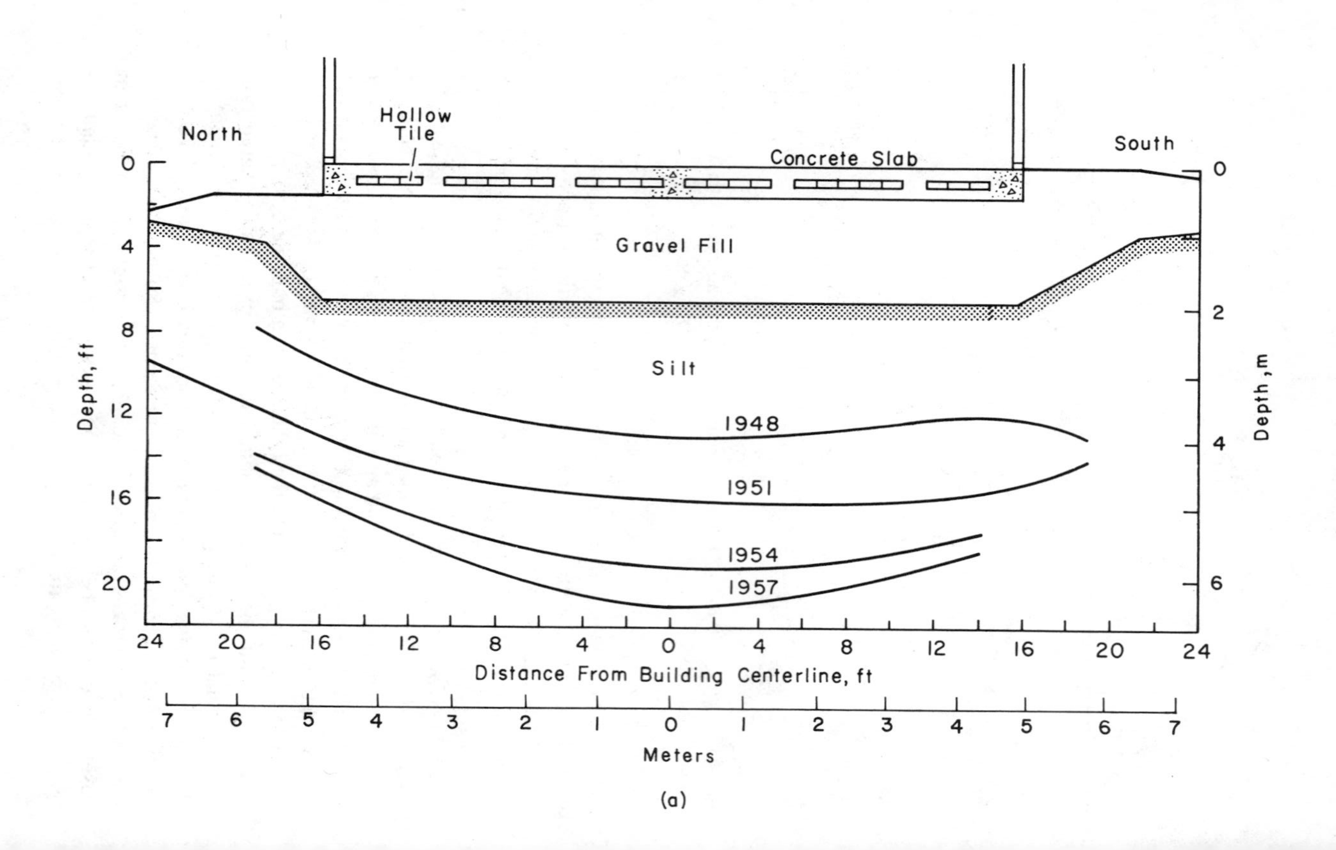
North
South
Hollow Tile
Concrete Slab
Gravel Fill
Silt
1948
1951
1954
1957
Depth, ft
0
4
8
12
16
20
Depth, m
0
2
4
6
24
20
16
12
8
4
0
4
8
12
16
20
24
Distance From Building Centerline, ft
7
6
5
4
3
2
1
0
1
2
3
4
5
6
7
Meters

(a)

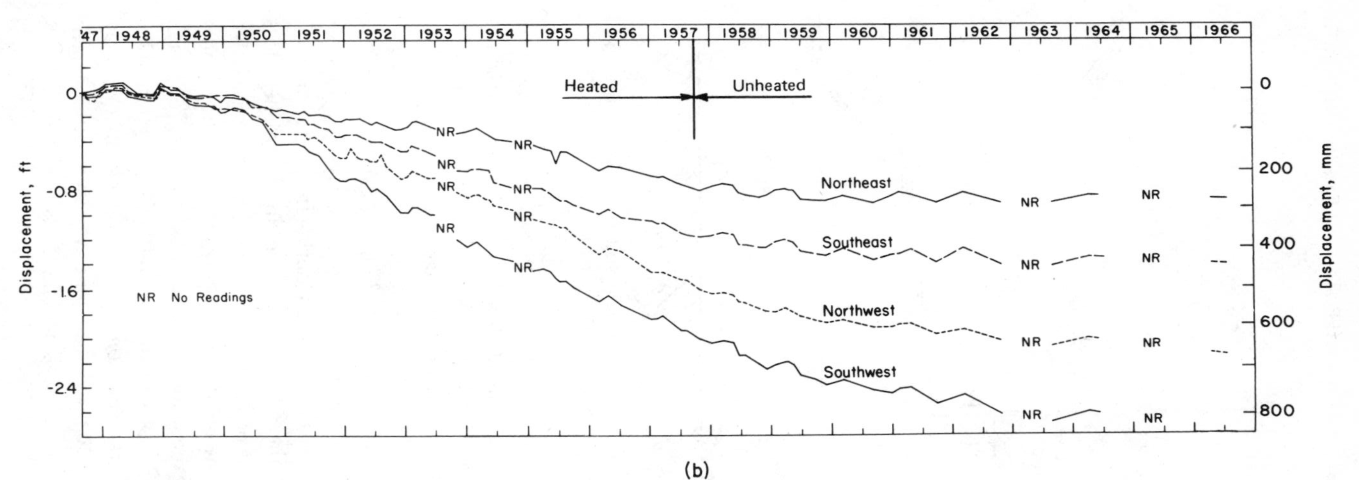

(b)

Fig. 11.4. Degradation of permafrost under a 32 x 32-ft (9.75 x 9.75 m) wood frame garage on uninsulated and inadequately ventilated concrete raft foundation, Fairbanks, Alaska: (a) degradation of permafrost on N-S centerline, 1948–1957; (b) displacement of building corners, 1947–1966 (Linell, Lobacz, et al. 1980, extension of data analyzed by Lobacz and Quinn 1966 with Kersten).

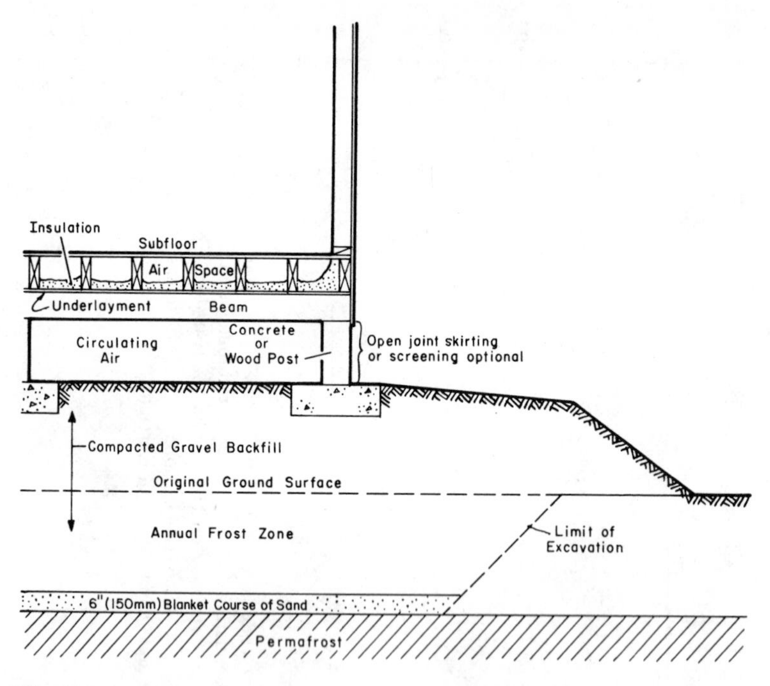

Fig. 11.5. Typical footing-type foundation for light structure with air space and gravel mat (Linell, Lobacz, et al. 1980).

however, the result will be settlement of overlying soil and structure supported thereon, as illustrated in Fig. 11.4(b). Such settlement is commonly differential and can cause extreme damage to conventional structures.

Degradation of permafrost under a heated building can be controlled by use of a ventilated foundation, which allows the layer of summer thaw in the foundation to refreeze completely during the winter, in combination with insulation of the floor to minimize downward heat loss from the building. An example of a simple ventilated foundation is shown in Fig. 11.5. In the design, the building also serves to shade the foundation from solar radiation (except for a few feet at the edges, for which supplementary shading can be provided), thus further helping to reduce thaw penetration. A mat of granular, non-frost-susceptible material may be used as shown in Fig. 11.5 to provide an upper soil layer within which freeze and thaw can occur without detrimental effects.

Warm pipe systems such as water, sewer or steam pipes, utilidors, or oil pipelines can produce severe effects in permafrost which are similar to those produced by heat-emitting buildings except with respect to geometry. The thaw zone tends to be cylindrical. In thaw-unstable permafrost, such thaw may cause overstressing and rupture of the pipe from differential settlement, buckling of the pipe on slopes from anchorage failure, soil erosion, and other adverse environmental effects. Thermal insulation of underground pipes can reduce but not entirely prevent heat loss into the surrounding ground, and great difficulty may be experienced in attempting to maintain insulation effectiveness under moist underground conditions. Therefore pipes are usually carried above ground under difficult permafrost conditions. Lines which carry fluids or gases at below-freezing temperatures do not cause thawing of permafrost, but if laid in frost-susceptible non-permafrost materials may experience differential frost heaving as the surrounding soil gradually freezes.

12. Engineering soil classification systems

Introduction

Classification provides an orderly and systematic way of identifying and describing soils for scientific or engineering purposes. For an engineer or scientist to apply the basic knowledge discussed in the preceding chapters—concerning thermal effects, frost heave, thaw settlement, ground moisture effects, and the general properties of soil, rock, and ice —he must be supplied with specific information by the field team and supporting laboratory, identifying and describing the materials present at the site, giving their characteristics to the extent needed, and detailing their horizontal and vertical distributions. Soil classification helps in the presentation of such information and also makes easier such tasks as selecting locations for detailed subsurface explorations and selecting samples for the more complex design tests. Nevertheless, classification should never be considered more than a step in the geotechnical design process, one that furnishes the engineer a general rather than a specific indication of behavioral characteristics. Design usually requires a more detailed evaluation of the soils than is provided by the classification system alone.

Depending upon the ultimate use of the information, the soil may be identified in accordance with one or more of several classification schemes. Regardless of the system used, a common basis should be used by field, laboratory, and all user personnel so that the characteristics and behavior will be accurately conveyed and understood by all. Using a widely accepted system allows the information to be readily reviewed by others and to be used by professional people at large when it is published in national or international scientific or technical journals. Soils may be classified with the aid of laboratory tests or solely on the basis of field identification. The latter method is especially valuable in the reconnaissance and preliminary design stages of construction projects when a specific site is still being selected. It is also valuable to the field inspection staff during the construction phases of projects.

The two most important engineering soil classification systems for use in arctic and subarctic areas are the Unified Soil Classification System and the AASHTO Classification System.

The unified soil classification system

The Unified System is a modification of the original Casagrande Airfield Classification System. In presenting his system, Casagrande (1948) included a thorough review and comparison of other existing soil classifications used in civil engineering. The Unified System was adopted by the U.S. Army Corps of Engineers and the U.S. Bureau of Reclamation in January 1952 and has been described in their publications (U.S. Army Corps of Engineers 1953; U.S. Bureau of Reclamation 1963). In 1959 the system was adopted by the U.S. Federal Housing Administration for use in residential developments.

The Unified System has been widely accepted both in the United States and internationally, and it has been used successfully for all types of construction. A supplementary system to cover classification of frozen soils has been published by Linell and Kaplar (1966). It became a part of the U.S. Defense Department Military Standard Unified Soil Classification System for Roads, Embankments and Foundations, mandatory for use by all departments of the U.S. Department of Defense, in 1968. The system for describing and classifying frozen soils represents the joint efforts of representatives of the Division of Building Research of the National Research Council of Canada, and the Arctic Construction and Frost Effects Laboratory of the U.S. Army Corps of Engineers. The basic elements of this system have been published separately by the National Research Council of Canada (Pihlainen and Johnston 1963) in a guide to field description of permafrost.

The Unified System is based on identifying soils according to the texture and plasticity of their ingredients and on grouping soils with respect to their behavior in a remolded or reworked condition. That the classification is for the remolded condition does not prevent its use for undisturbed soil. The following properties form the basis for dividing soils into 15 groups:

Percentages of gravel, sand, and fines (fraction passing No. 200 [0.075 mm] sieve)
Shape of the grain size curve
Plasticity and compressibility characteristics

The following letter designations are used to compose symbols for the groups:

G = Gravel
S = Sand

M = Inorganic silt or very fine sand (nonplastic or low plasticity fines)
C = Clay (primarily inorganic)
Pt = Peat and other highly organic soils
O = Organic
W = Well-graded
P = Poorly graded
L = Low compressibility
H = High compressibility

In the Unified System, the limiting boundaries between the various soil size ranges have been set at certain U.S. Standard sieve sizes as shown in Table 12.1.

Soil having 50 percent or less passing the No. 200 (0.075 mm) sieve is termed *coarse-grained,* and soil having more than 50 percent passing the No. 200 (0.075 mm) sieve is termed *fine-grained.* Coarse-grained soils are assigned the symbol G if more than 50 percent of the part coarser than the No. 200 (0.075 mm) sieve size is larger than the No. 4 (4.75 mm) sieve, and the symbol S if more than 50 percent is smaller than the No. 4 (4.75 mm) sieve size. The G or S is followed by W, P, M, or C to denote the gradation or the amount and kind of fines present. See columns 2 and 7 of Table 12.2. Plasticity characteristics are determined on the minus No. 40 (0.425 mm) sieve size fraction. Well-graded coarse-grained materials with less than 5 percent passing the No. 200 (0.075 mm) sieve and which are free-draining and lack plasticity are classified as GW or SW; poorly graded coarse-grained materials meeting similar criteria are classified as GP or SP. Well-graded materials generally have smooth and regular concave curves with no sizes lacking and no excess of material in any one size range, as indicated by the GW and SW curves in Fig. 12.1. The GP and SP curves in Fig. 12.1 are examples of poorly graded soils. To be classed as well-graded, materials should meet the criteria for uniformity coefficient, C_u, and coefficient of curvature, C_c, shown in column 7 of Table 12.2 opposite the Gravel and Sand Categories, respectively. The M or C designations for G and S soils are determined by where the liquid limit and plasticity index of the minus No. 40 (0.425 mm) sieve size fraction plot above or below the A line in the chart at the lower right of Table 12.2. The gradation of coarse-grained materials carrying the M or C designations is not considered significant and no distinction is made between well-graded and poorly graded materials in these categories. For soils with more than 12 percent passing the No. 200 (0.075 mm) sieve, single group designations (GM, SM, GC, SC) are assigned. Soils with between 5 and 12 percent passing the No.

TABLE 12.1.
Size classification of soil particles (in mm)

System or agency [a]	Cobbles and boulders	Gravel		Very coarse sand	Coarse sand	Medium sand	Fine sand	Very fine sand	Silt	Clay
	Cobbles	Coarse gravel	Fine gravel						Fines (silt or clay)	
Unified classification system	>75 (3″)	75–19	19–4.75	....	4.75–2.0	2.0–0.425	0.425–0.075	....	<0.075	
	Boulders								Silt-clay	
AASHTO classification system	>75	75-2.0		....	2.0–0.425	....	0.425–0.075	....	<0.075	
USDA		>2.0		2.0–1.0	1.0–0.5	0.5–0.25	0.25–0.10	0.10–0.05	0.05–0.002	<0.002
MIT		>2.0		....	2.0–0.6	0.06–0.2	0.2–0.06	....	0.06–0.002	<0.002
ISSS		>2.0		....	2.0–0.2	....	0.2–0.02	....	0.02–.0002	<0.002

[a] AASHTO—American Association of State Highway and Transportation Officials, Designation M145-73.
USDA—U.S. Department of Agriculture.
MIT—Massachusetts Institute of Technology.
ISSS—International Society of Soil Science.
Source: Adapted in part from Engineering characteristics of soil and soil testing, by M. G. Spangler, in *Highway engineering handbook* (K. B. Woods, ed.), copyright © 1960 by the McGraw-Hill Book Company, Inc., and used with permission of the McGraw-Hill Book Company.

TABLE 12.2.

Unified soil classification system

(including identification and description)

UNIFIED SOIL CLASSIFICATION
(Including Identification and Description)

Major Divisions (1)	(2)	(2)	Group Symbols (3)	Typical Names (4)	Field Identification Procedures (Excluding particles larger than 3 inches and basing fractions on estimated weights) (5)	Information Required for Describing Soils (6)	Laboratory Classification Criteria (7)
Coarse-grained Soils More than half of material is larger than No. 200 sieve size. The No. 200 sieve size is about the smallest particle visible to the naked eye.	Gravels More than half of coarse fraction is larger than No. 4 sieve size. (For visual classification, the 1/4-in. size may be used as equivalent to the No. 4 sieve size)	Clean Gravels (Little or no fines)	GW	Well-graded gravels, gravel-sand mixtures, little or no fines.	Wide range in grain sizes and substantial amounts of all intermediate particle sizes.	For undisturbed soils add information on stratification, degree of compactness, cementation, moisture conditions and drainage characteristics.	$C_u = \frac{D_{60}}{D_{10}}$ Greater than 6; $C_c = \frac{(D_{30})^2}{D_{10} \times D_{60}}$ Between one and 3
			GP	Poorly-graded gravels, gravel-sand mixtures, little or no fines.	Predominantly one size or a range of sizes with some intermediate sizes missing.		Not meeting all gradation requirements for GW
		Gravels with Fines (Appreciable amount of fines)	GM	Silty gravels, gravel-sand-silt mixtures.	Nonplastic fines or fines with low plasticity (for identification procedures see ML below).	Give typical name; indicate approximate percentages of sand and gravel, max. size; angularity, surface condition, and hardness of the coarse grains; local or geologic name and other pertinent descriptive information; and symbol in parentheses.	Atterberg limits below "A" line or PI less than 4. Above "A" line with PI between 4 and 7 are borderline cases requiring use of dual symbols.
			GC	Clayey gravels, gravel-sand-clay mixtures.	Plastic fines (for identification procedures see CL below).		Atterberg limits above "A" line with PI greater than 7
	Sands More than half of coarse fraction is smaller than No. 4 sieve size.	Clean Sands (Little or no fines)	SW	Well-graded sands, gravelly sands, little or no fines.	Wide range in grain size and substantial amounts of all intermediate particle sizes.		$C_u = \frac{D_{60}}{D_{10}}$ Greater than 4; $C_c = \frac{(D_{30})^2}{D_{10} \times D_{60}}$ Between one and 3
			SP	Poorly-graded sands, gravelly sands, little or no fines.	Predominantly one size or a range of sizes with some intermediate sizes missing.		Not meeting all gradation requirements for SW
		Sands with Fines (Appreciable amount of fines)	SM	Silty sands, sand-silt mixtures.	Nonplastic fines or fines with low plasticity (for identification procedures see ML below).	Example: Silty sand, gravelly; about 20% hard, angular gravel particles 1/2-in. maximum size; rounded and subangular sand grains coarse to fine; about 15% nonplastic fines with low dry strength; well compacted and moist in place; alluvial sand; (SM).	Atterberg limits below "A" line or PI less than 4. Limits plotting in hatched zone with PI between 4 and 7 are borderline cases requiring use of dual symbols.
			SC	Clayey sands, sand-clay mixtures.	Plastic fines (for identification procedures see CL below).		Atterberg limits above "A" line with PI greater than 7

Use grain-size curve in identifying the fractions as given under field identification.

Determine percentages of gravel and sand from grain-size curve. Depending on percentage of fines (fraction smaller than No. 200 sieve size) coarse-grained soils are classified as follows:
Less than 5% GW, GP, SW, SP,
More than 12% GM, GC, SM, SC.
5% to 12% Borderline cases requiring use of dual symbols.

Major Divisions (1)	(2)	Group Symbols (3)	Typical Names (4)	Identification Procedures on Fraction Smaller than No. 40 Sieve Size: Dry Strength (Crushing characteristics)	Dilatancy (Reaction to shaking)	Toughness (Consistency near PL)	Information Required for Describing Soils (6)
Fine-grained Soils More than half of material is smaller than No. 200 sieve size.	Silts and Clays Liquid limit less than 50	ML	Inorganic silts and very fine sands, rock flour, silty or clayey fine sands or clayey silts with slight plasticity.	None to slight	Quick to slow	None	Give typical name, indicate degree and character of plasticity, amount and maximum size of coarse grains, color in wet condition, odor if any, local or geologic name, and other pertinent descriptive information; and symbol in parentheses. 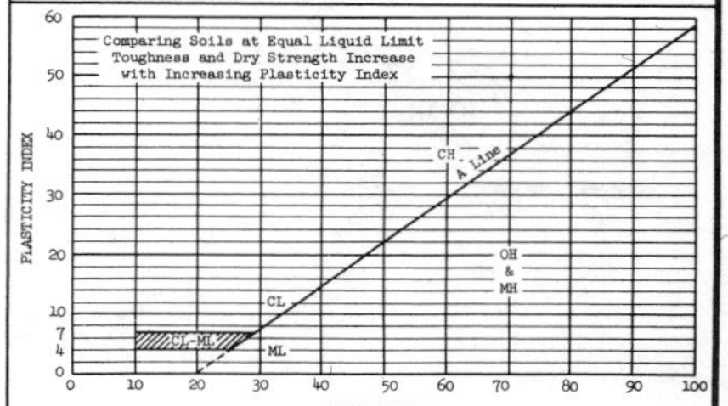
		CL	Inorganic clays of low to medium plasticity, gravelly clays, sandy clays, silty clays, lean clays.	Medium to high	None to very slow	Medium	
		OL	Organic silts and organic silty clays of low plasticity.	Slight to medium	Slow	Slight	
	Silts and Clays Liquid limit greater than 50	MH	Inorganic silts, micaceous or diatomaceous fine sandy or silty soils, elastic silts.	Slight to medium	Slow to none	Slight to medium	For undisturbed soils add information on structure, stratification, consistency in undisturbed and remolded states, moisture and drainage conditions.
		CH	Inorganic clays of high plasticity, fat clays.	High to very high	None	High	
		OH	Organic clays of medium to high plasticity, organic silts.	Medium to high	None to very slow	Slight to medium	Example: Clayey silt, brown, slightly plastic, small percentage of fine sand, numerous vertical root holes, firm and dry in place, loess, (ML).
Highly Organic Soils		Pt	Peat and other highly organic soils.	Readily identified by color, odor, spongy feel and frequently by fibrous texture.			

PLASTICITY CHART

For laboratory classification of fine-grained soils

(1) Boundary classifications: Soils possessing characteristics of two groups are designated by combinations of group symbols. For example GW-GC, well-graded gravel-sand mixture with clay binder. (2) All sieve sizes on this chart are U. S. standard.

Source: *The Unified Soil Classification System,* U.S. Army Corps of Engineers 1953.

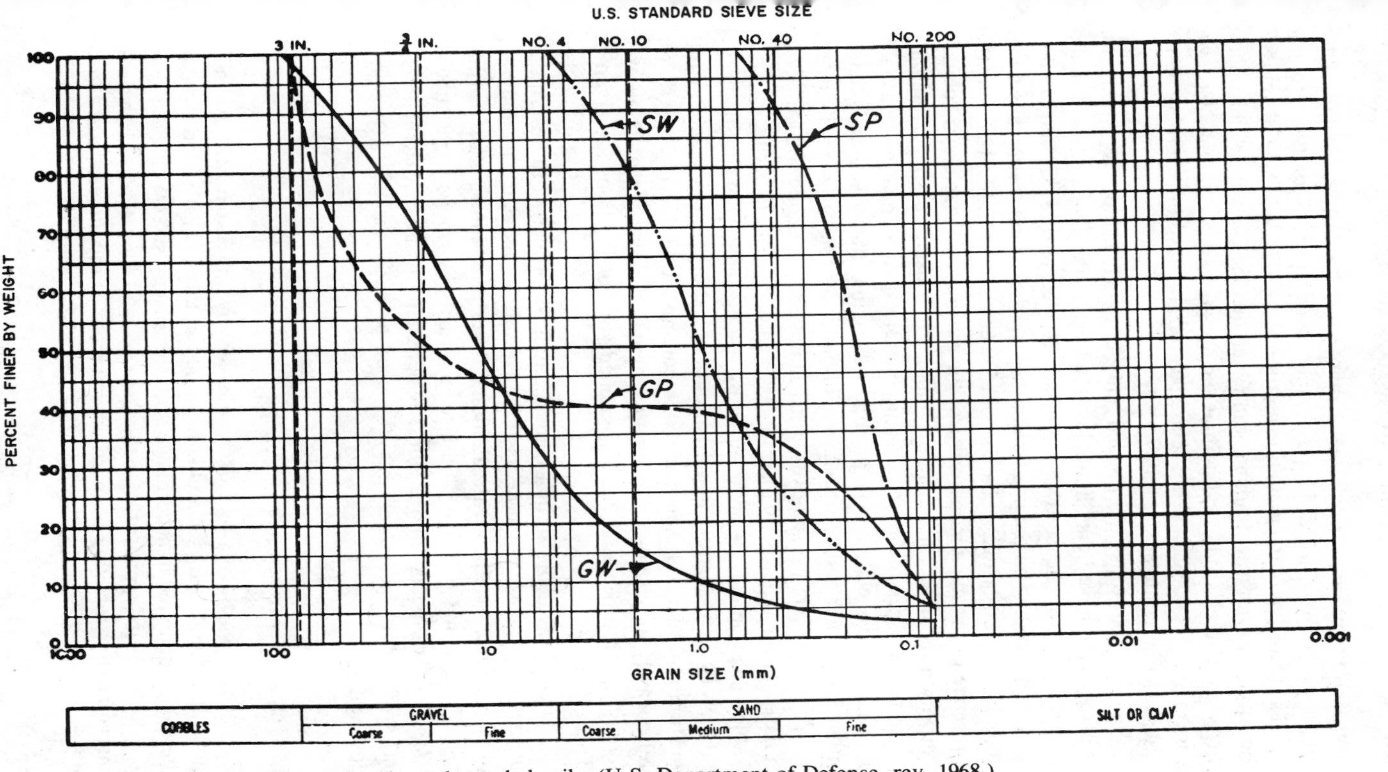

Fig. 12.1. Typical well-graded and poorly graded soils. (U.S. Department of Defense, rev. 1968.)

200 (0.075 mm) sieve are classed as borderline and require dual designations, e.g., GW–GM. Soils possessing characteristics of two groups, such as soils that fit the gradation requirements for GW, SW, GP, and SP but which are not free-draining or in which the fine fraction exhibits plasticity are also classed as borderline and are given dual symbols. Criteria for GM, GC, SM, SC, and dual classifications on basis of Atterberg limits are shown in column 7 of Table 12.2.

Fine-grained soils are divided into L or H groups depending on whether they have liquid limits less than or greater than 50 (see column 2 of Table 12.2). These two groups are divided into M, C, and O categories, depending on the position of the Atterberg limit test results when plotted on the plasticity chart at the lower right of Table 12.2 and the presence of organic characteristics; the various applicable symbols are shown on the chart. A separate symbol, Pt, is used for soils containing large percentages of fibrous organic matter, such as peat and partially decomposed vegetation. The presence of organic matter is usually determinable by color and odor, but in doubtful cases the Atterberg limit test can be made on moist and oven-dried samples. An organic soil will show a radical drop in plasticity after either oven-drying or air-drying.

Identification may be made either by testing samples of the soil in the laboratory for gradation, liquid limit, and plastic limit or by field identification procedures as summarized in column 5 of Table 12.2; the latter rely heavily on visual and manual examination. Further details on field identification procedures for fine-grained soils or fractions are given in Table 12.3.

The group symbols are not a substitute for a proper word description of the soil. Names should be used for the soils as indicated in column 4 of Table 12.2, together with additional descriptive information as indicated in column 6 of Table 12.2. Use of special descriptive terms such as *Glacial Till* or locally coined words which help to convey the characteristics of the soil accurately may be used as *additions* to the required description.

Classification of frozen soils

The system for classifying frozen soils is shown in Table 12.4. As indicated in column 1, frozen soil is identified in three stages. Under part I the soil is identified in accordance with the basic Unified Soil Classification System, independently of the frozen state. Under part II the soil characteristics resulting from the frozen state of the material are added

TABLE 12.3.
Field identification procedures for fine-grained soils or fractions

These procedures are to be performed on the minus No. 40 sieve size (0.425 mm) particles, approximately 1/64 in. For field classification purposes, screening is not intended, simply remove by hand the coarse particles that interfere with the tests.

Dilatancy (reaction to shaking)

After removing particles larger than No. 40 sieve size, prepare a pat of moist soil with a volume of about one-half cubic inch (8×10^3 mm^3). Add enough water if necessary to make the soil soft but not sticky.

Place the pat in the open palm of one hand and shake horizontally, striking vigorously against the other hand several times. A positive reaction consists of the appearance of water on the surface of the pat which changes to a livery consistency and becomes glossy. When the sample is squeezed between the fingers, the water and gloss disappear from the surface, the pat stiffens, and finally it cracks or crumbles. The rapidity of appearance of water during shaking and of its disappearance during squeezing assist in identifying the character of the fines in a soil.

Very fine clean sands give the quickest and most distinct reaction whereas a plastic clay has no reaction. Inorganic silts, such as a typical rock flour, show a moderately quick reaction.

Dry Strength (crushing characteristics)

After removing particles larger than No. 40 sieve size, mold a pat of soil to the consistency of putty, adding water if necessary. Allow the pat to dry completely by oven, sun, or air-drying, and then test its strength by breaking and crumbling between the fingers. This strength is a measure of the character and quantity of the colloidal fraction contained in the soil. The dry strength increases with increasing plasticity.

High dry strength is characteristic for clays of the CH group. A typical inorganic silt possesses only very slight dry strength. Silty fine sands and silts have about the same slight dry strength, but can be distinguished by the feel when powdering the dried specimen. Fine sand feels gritty whereas a typical silt has the smooth feel of flour.

Toughness (consistency near plastic limit)

After particles larger than the No. 40 sieve size are removed, a specimen of soil about one-half inch (13 mm) cube in size, is molded to the consistency of putty. If too dry, water must be added and if sticky, the specimen should be spread out in a thin layer and allowed to lose some moisture by evaporation. Then the specimen is rolled out by hand on a smooth surface or between the palms into a thread about one-eighth in. (3 mm) diameter. The thread is then folded and rerolled repeatedly. During this manipulation the moisture content is gradually reduced and the specimen stiffens, finally loses its plasticity, and crumbles when the plastic limit is reached.

After the thread crumbles, the pieces should be lumped together and a slight kneading action continued until the lump crumbles.

The tougher the thread near the plastic limit and the stiffer the lump when it finally crumbles, the more potent is the colloidal clay fraction in the soil. Weakness of the thread at the plastic limit and quick loss of coherence of the lump below the plastic limit indicate either inorganic clay of low plasticity, or materials such as kaolin-type clays and organic clays which occur below the A-line.

Highly organic clays have a very weak and spongy feel at the plastic limit.

Source: *The Unified Soil Classification System,* U.S. Army Corps of Engineers 1953.

TABLE 12.4.
Description and classification of frozen soils

(1)	Major Group: Description (2)	Major Group: Designation (3)	Sub-Group: Description (4)	Sub-Group: Designation (5)	Field Identification (6)	Pertinent Properties of Frozen Materials Which May be Measured by Physical Tests to Supplement Field Identification (7)	Guide for Construction on Soils Subject to Freezing and Thawing: Thaw Characteristics (8)	Guide for Construction on Soils Subject to Freezing and Thawing: Criteria (9)
PART I **DESCRIPTION OF SOIL PHASE (a)** (Independent of Frozen State)	Classify Soil Phase by the Unified Soil Classification System							
PART II **DESCRIPTION OF FROZEN SOIL**	Segregated ice is not visible by eye (b)	N	Poorly bonded or friable	Nf	Identify by visual examination. To determine presence of excess ice, use procedure under note (c) below and hand magnifying lens as necessary. For soils not fully saturated, estimate degree of ice saturation: Medium, Low. Note presence of crystals, or of ice coatings around larger particles.	In-Place Temperature Density and Void Ratio a. In Frozen State b. After Thawing in Place Water Content (total H_2O, including ice) a. Average b. Distribution Strength a. Compressive b. Tensile c. Shear d. Adfreeze Elastic Properties Plastic Properties Thermal Properties Ice Crystal Structure (using optical instruments) a. Orientation of Axes b. Crystal Size c. Crystal Shape d. Pattern of Arrangement	Usually thaw-stable	The potential intensity of ice segregation in a soil is dependent to a large degree on its void sizes and for pavement design purposes may be expressed as an empirical function of grain size as follows: Most inorganic soils containing 3 percent or more of grains finer than 0.02 mm in diameter by weight are frost-susceptible for pavement design purposes. Gravels, well-graded sands and silty sands, especially those approaching the theoretical maximum density curve, which contain 1½ to 3 percent finer by weight than 0.02 mm size should be considered as possibly frost-susceptible and should be subjected to a standard laboratory frost susceptibility test to evaluate actual behavior during freezing. Uniform sandy soils may have as high as 10 percent of grains finer than 0.02 mm by weight without being frost-susceptible. However, their tendency to occur interbedded with other soils usually makes it impractical to consider them separately. Soils classed as frost-susceptible under the above pavement design criteria are likely to develop significant ice segregation and frost heave if frozen at normal rates with free water readily available. Soils so frozen will fall into the thaw-unstable category. However, they may also be classed as thaw-stable if frozen with insufficient water to permit ice segregation. Soils classed as non-frost-susceptible (*NFS) under the above criteria usually occur without significant ice segregation and are usually thaw-stable for pavement applications. However, the criteria are not exact and may be inadequate for some structure applications: exceptions may also result from minor soil variations. In permafrost areas, ice wedges, pockets, veins, or other ice bodies may be found whose mode of origin is different from that described above. Such ice may be the result of long-time surface expansion and contraction phenomena or may be glacial or other ice which has been buried under a protective earth cover.
			Well bonded: No excess ice	Nb: n				
			Well bonded: Excess ice	Nb: e				
	Segregated ice is visible by eye. (Ice 1 inch or less in thickness) (b)	V	Individual ice crystals or inclusions	Vx	For ice phase, record the following as applicable: Location, Orientation, Thickness, Length, Spacing, Hardness, Structure, Color (Hardness, Structure, Color per Part III below); Size, Shape, Pattern of arrangement Estimate volume of visible segregated ice present as percent of total sample volume.		Usually thaw-unstable	
			Ice coatings on particles	Vc				
			Random or irregularly oriented ice formations	Vr				
			Stratified or distinctly oriented ice formations	Vs				
PART III **DESCRIPTION OF SUBSTANTIAL ICE STRATA**	Ice (Greater than 1 inch in thickness)	ICE	Ice with soil inclusions	ICE + soil type	Designate material as ICE (d) and use descriptive terms as follows, usually one item from each group, as applicable: Hardness: HARD, SOFT (of mass, not individual crystals) Structure: CLEAR, CLOUDY, POROUS, CANDLED, GRANULAR, STRATIFIED Color: (Examples): COLORLESS, GRAY, BLUE Admixtures: (Example): CONTAINS FEW THIN SILT INCLUSIONS	Same as Part II above, as applicable, with special emphasis on Ice Crystal Structure.		
			Ice without soil inclusions	ICE				

DEFINITIONS:

Ice Coatings on Particles are discernible layers of ice found on or below the larger soil particles in a frozen soil mass. They are sometimes associated with hoarfrost crystals, which have grown into voids produced by the freezing action.

Ice Crystal is a very small individual ice particle visible in the face of a soil mass. Crystals may be present alone or in a combination with other ice formations.

Clear Ice is transparent and contains only a moderate number of air bubbles. (e)

Cloudy Ice is translucent, but essentially sound and non-pervious. (e)

Porous Ice contains numerous voids, usually interconnected and usually resulting from melting at air bubbles or along crystal interfaces from presence of salt or other materials in the water, or from the freezing of saturated snow. Though porous, the mass retains its structural unity.

Candled Ice is ice which has rotted or otherwise formed into long columnar crystals, very loosely bonded together.

Granular Ice is composed of coarse, more or less equidimensional, ice crystals weakly bonded together.

Ice Lenses are lenticular ice formations in soil occurring essentially parallel to each other, generally normal to the direction of heat loss and commonly in repeated layers.

Ice Segregation is the growth of ice as distinct lenses, layers, veins, and masses in soils, commonly but not always oriented normal to direction of heat loss.

Well-bonded signifies that the soil particles are strongly held together by the ice and that the frozen soil possesses relatively high resistance to chipping or breaking.

Poorly-bonded signifies that the soil particles are weakly held together by the ice and that the frozen soil consequently has poor resistance to chipping or breaking.

Friable denotes a condition in which material is easily broken up under light to moderate pressure.

Thaw-Stable frozen soils do not, on thawing, show loss of strength below normal, long-time thawed values nor produce detrimental settlement.

Thaw-Unstable frozen soils show, on thawing, significant loss of strength below normal, long-time thawed values and/or significant settlement, as a direct result of the melting of the excess ice in the soil.

NOTES:

(a) When rock is encountered, standard rock classification terminology should be used.

(b) Frozen soils in the N group may, on close examination, indicate presence of ice within the voids of the material by crystalline reflections or by a sheen on fractured or trimmed surfaces. However, the impression to the unaided eye is that none of the frozen water occupies space in excess of the original voids in the soil. The opposite is true of frozen soils in the V group.

(c) When visual methods may be inadequate, a simple field test to aid evaluation of volume of excess ice can be made by placing some frozen soil in a small jar, allowing it to melt and observing the quantity of supernatant water as a percent of total volume.

(d) Where special forms of ice, such as hoarfrost, can be distinguished, more explicit description should be given.

(e) Observer should be careful to avoid being misled by surface scratches or frost coating on the ice.

NOTES:

The letter symbols shown are to be affixed to the Unified Soil Classification letter designations, or may be used in conjunction with graphic symbols, in exploration logs or geological profiles. Example - a lean clay with essentially horizontal ice lenses.

CL-Vs or Vs

The descriptive name of the frozen soil type and a complete description of the frozen material are the fundamental elements of this classification scheme. Additional descriptive data should be added where necessary. The letter symbols are secondary and are intended only for convenience in preparing graphical presentations. Since it is frequently impractical to describe ice formations in frozen soils by means of words alone, sketches and photographs should be used where appropriate, to supplement descriptions.

*The abbreviation NFS is commonly used to designate non-frost-susceptible materials on exploration logs and drawings.

to the soil description. Under part III important ice strata present in the soil are described.

As shown in columns 2 and 3 of Table 12.4, under part II, frozen soils are divided into two major groups: soils in which segregated ice is *not* visible to the unaided eye (designation N), and soils in which segregated ice *is visible* (designation V). The latter grouping is applied only to soil containing ice layers 1 in. or less in thickness. As shown in columns 4 and 5 of Table 12.4, soils in which segregated ice is not visible to the eye are divided into friable (Nf) and well-bonded (Nb) frozen soil, and the latter is subdivided into soils without excess ice (Nbn) and with (Nbe). Frozen soils in which ice is visible to the naked eye are divided into the following four subgroups, arranged approximately in order of increasing ice content as commonly encountered:

Vx—individual ice crystals or inclusions
Vc—ice coatings on particles
Vr—random or irregularly oriented ice formations
Vs—stratified or distinctly oriented ice formations

As indicated in columns 2 and 3 of Table 12.4, under part III, substantial ice strata greater than 1 in. in thickness are designated separately as ICE. As shown in columns 4 and 5 of Table 12.4, part III, two broad categories are recognized:

ICE plus soil type (ice with soil inclusion)
ICE (ice without soil inclusions)

When more than one subgroup characteristic is present in the same material, multiple subgroup designations may be used, as Vs,r. Guidance for field identification and supplementary tests is given in columns 6 and 7 of Table 12.4. Photographs of representative frozen soil types are shown in Figs. 12.2 and 12.3.

Graphical presentation of frozen soil data

The recommended procedure for graphical presentation of frozen soil classification data consists of showing the applicable symbols for the material type and its frozen condition, followed by appropriate descriptive material for both the soil and included ice features, as indicated in Fig. 12.4. Ice strata may be similarly identified and described. To simplify identification of the strata that are frozen, a wide bar may be drawn at the side of the graphic log, as indicated on Fig. 12.4.

Table 12.5 illustrates the general characteristics of the unfrozen soil groups of the Unified Soil Classification System in relation to roads and

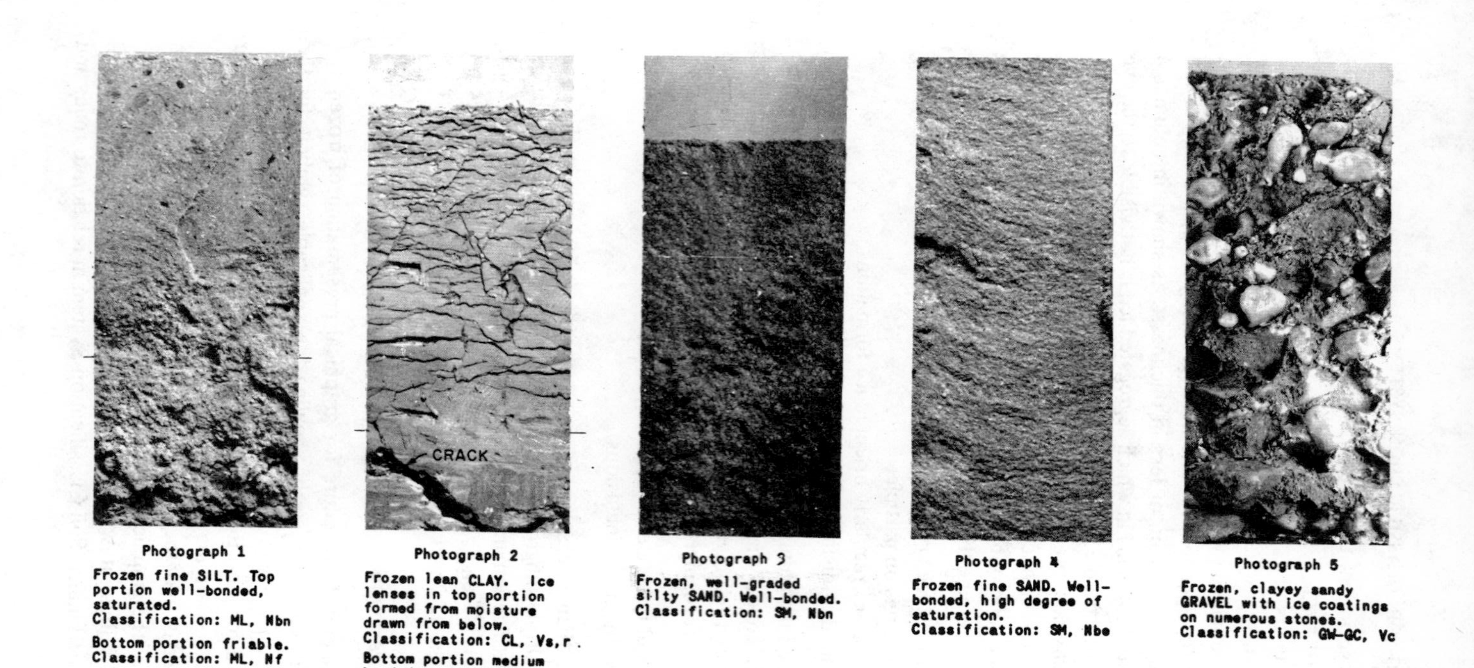

Fig. 12.2. Photographs of frozen soil types (Linell and Kaplar 1966. Adapted from *Permafrost: Proceedings of an International Conference,* p. 485, with the permission of the National Academy of Sciences, Washington, D.C.).

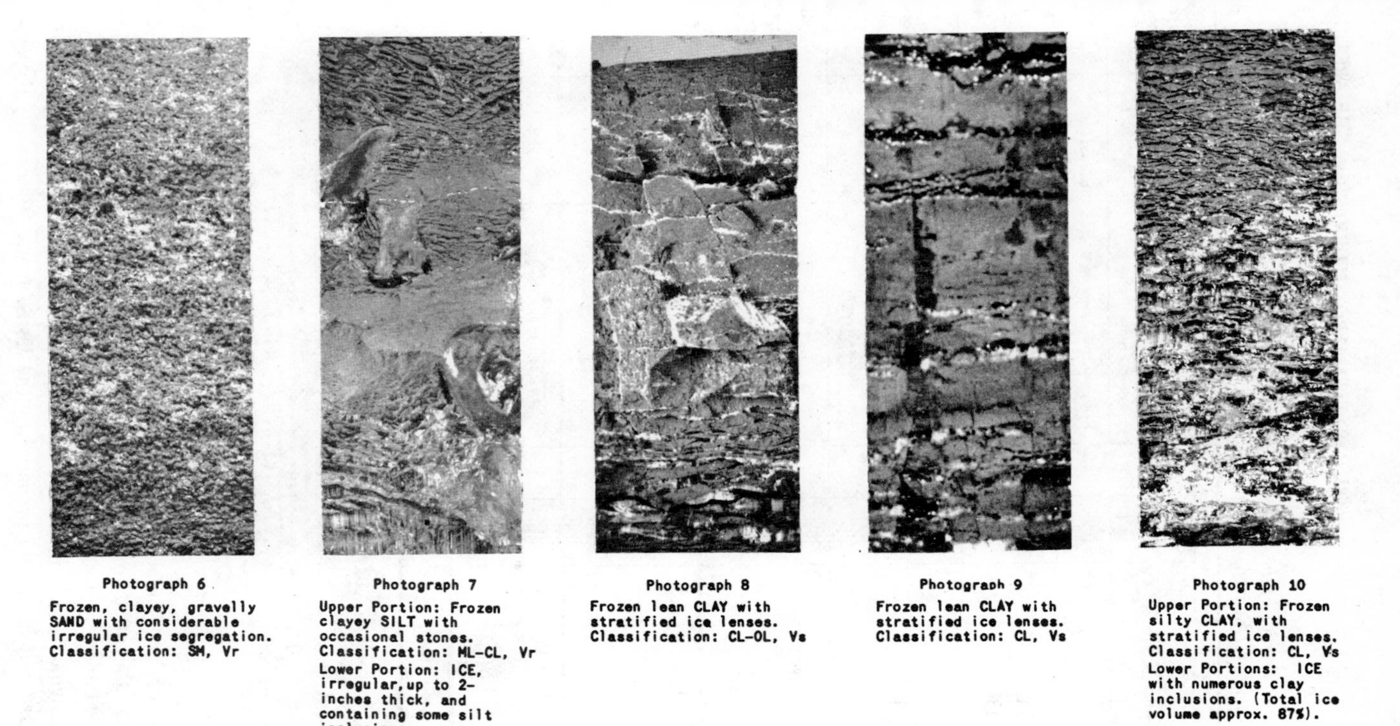

Fig. 12.3. Photographs of frozen soil types (Linell and Kaplar 1966. Adapted from *Permafrost: Proceedings of an International Conference*, p. 485, with the permission of the National Academy of Sciences, Washington, D.C.).

TABLE 12.5.

Characteristics of the Unified Soil Classification System pertinent to roads and airfields (U.S. Department of Defense, rev. 1968).

Major divisions (1)	(2)	Symbol: Letter (3)	Hatching (4)	Color (5)	Soil type (6)	Performance value as subgrade when not subject to frost action (7)	Performance v[alue] as subbase wh[en not] subject to frost [action] (8)
COARSE-GRAINED SOILS	GRAVEL AND GRAVELLY SOILS	GW		Red	Well-graded gravels or gravel-sand mixtures, little or no fines	Excellent	Excellent
		GP		Red	Poorly graded gravels or gravel-sand mixture, little or no fines	Good to excellent	Good
		GM d		Yellow	Silty gravels, gravel-sand-silt mixtures	Good to excellent	Good
		GM u		Yellow		Good	Fair
		GC		Yellow	Clayey gravels, gravel-sand-clay mixture	Good	Fair
	SAND AND SANDY SOILS	SW		Red	Well-graded sands or gravelly sands, little or no fines	Good	Fair to good
		SP		Red	Poorly graded sands or gravelly sands, little or no fines	Fair to good	Fair
		SM d		Yellow	Silty sands, sand-silt mixtures	Fair to good	Fair to good
		SM u		Yellow		Fair	Poor to fair
		SC		Yellow	Clayey sands, sand-clay mixtures	Poor to fair	Poor
FINE-GRAINED SOILS	SILTS AND CLAYS LL IS LESS THAN 50	ML		Green	Inorganic silts and very fine sands, rock flour, silty or clayey fine sands or clayey silts with slight plasticity	Poor to fair	Not suitable
		CL		Green	Inorganic clays of low to medium plasticity, gravelly clays, sandy clays, silty clays, lean clays	Poor to fair	Not suitable
		OL		Green	Organic silts and organic silt-clays of low plasticity	Poor	Not suitable
	SILTS AND CLAYS LL IS GREATER THAN 50	MH		Blue	Inorganic silts, micaceous or diatomaceous fine sandy or silty soils, elastic silts	Poor	Not suitable
		CH		Blue	Inorganic clays of high plasticity, fat clays	Poor to fair	Not suitable
		OH		Blue	Organic clays of medium to high plasticity, organic silts	Poor to very poor	Not suitable
HIGHLY ORGANIC SOILS		Pt		Orange	Peat and other highly organic soils	Not suitable	Not suitable

NOTE:

1. Column 3, division of GM and SM groups into subdivisions of d and u are for roads and airfields only. Subdivision is on basis of Atterberg limits; suffix d (e.g., GMd) will be used when the liquid limit is 25 or less and the plasticity index is 5 or less; the suffix u will be used otherwise.
2. In column 13, the equipment listed will usually produce the required densities with a reasonable number of passes when moisture conditions and thickness of lift are properly controlled. In some instances, several types of equipment are listed because variable soil charactertistics within a given soil group may require different equipment. In some instances, a combination of two types may be necessary.
   a. Processed based materials and other angular materials. Steel-wheeled and rubber-tired rollers are recommended for hard, angular materials with limited fines or screenings. Rubber-tired equipment is recommended for softer materials subject to degradation.
   b. Finishing. Rubber-tired equipment is recommended for rolling during final shaping operations for most soils and processed materials.

airfields, and Table 12.5(a) shows characteristics of the soil groups pertinent to embankments and foundations. The relationships shown in these tables are discussed in considerable detail by the U.S. Army Corps of Engineers in Appendices A and B of the report *The Unified Soil Classification System* prepared by the U. S. Army Waterways Experi-

rmance value base when not ct to frost action (9)	Potential frost action (10)	Compressibility and expansion (11)	Drainage characteristics (12)	Compaction equipment (13)	Unit dry weight lb per cu ft (14)	Typical design values CBR (15)	Subgrade modulue k lb per cu in. (16)
	None to very slight	Almost none	Excellent	Crawler-type tractor, rubber-tired roller, steel-wheeled roller	125-140	40-80	300-500
good	None to very slight	Almost none	Excellent	Crawler-type tractor, rubber-tired roller, steel-wheeled roller	110-140	30-60	300-500
good	Slight to medium	Very slight	Fair to poor	Rubber-tired roller, sheepsfoot roller; close control of moisture	125-145	40-60	300-500
to not suitable	Slight to medium	Slight	Poor to practically impervious	Rubber-tired roller, sheepsfoot roller	115-135	20-30	200-500
to not suitable	Slight to medium	Slight	Poor to practically impervious	Rubber-tired roller, sheepsfoot roller	130-145	20-40	200-500
	None to very slight	Almost none	Excellent	Crawler-type tractor, rubber-tired roller	110-130	20-40	200-400
to not suitable	None to very slight	Almost none	Excellent	Crawler-type tractor, rubber-tired roller	105-135	10-40	150-400
	Slight to high	Very slight	Fair to poor	Rubber-tired roller, sheepsfoot roller; close control of moisture	120-135	15-40	150-400
suitable	Slight to high	Slight to medium	Poor to practically impervious	Rubber-tired roller, sheepsfoot roller	100-130	10-20	100-300
suitable	Slight to high	Slight to medium	Poor to practically impervious	Rubber-tired roller, sheepsfoot roller	100-135	5-20	100-300
suitable	Medium to very high	Slight to medium	Fair to poor	Rubber-tired roller, sheepsfoot roller; close control of moisture	90-130	15 or less	100-200
suitable	Medium to high	Medium	Practically impervious	Rubber-tired roller, sheepsfoot roller	90-130	15 or less	50-150
suitable	Medium to high	Medium to high	Poor	Rubber-tired roller, sheepsfoot roller	90-105	5 or less	50-100
suitable	Medium to very high	High	Fair to poor	Sheepsfoot roller, rubber-tired roller	80-105	10 or less	50-100
suitable	Medium	High	Practically impervious	Sheepsfoot roller, rubber-tired roller	90-115	15 or less	50-150
suitable	Medium	High	Practically impervious	Sheepsfoot roller, rubber-tired roller	80-110	5 or less	25-100
suitable	Slight	Very high	Fair to poor	Compaction not practical	—	—	—

c. Equipment size. The following sizes of equipment are necessary to assure the high densities required for airfield construction:

Crawler-type tractor — total weight in excess of 30,000 lb.

Rubber-tired equipment — wheel load in excess of 15,000 lb, wheel loads as high as 40,000 lb may be necessary to obtain the required densities for some materials (based on contact pressure of approximately 65 to 150 psi).

Sheepsfoot roller — unit pressure (on 6- to 12-sq. in. foot) to be in excess of 250 psi and unit pressures as high as 650 psi may be necessary to obtain the required densities for some materials. The area of the feet should be at least 5 per cent of the total peripheral area of the drum, using the diameter measured to the faces of the feet.

3. Column 14, unit dry weights are for compacted soil at optimum moisture conent for modified AASHTO compaction effort.
4. In column 15, the maximum value that can be used in design of airfields is, in some cases, limited by gradation and plasticity requirements.

Depth	Symbol	SOIL DESCRIPTION	ICE FEATURES
0.0*	OL	Organic, sandy SILT, not frozen	None
0.5	GW	Brown, well-graded, sandy GRAVEL, medium compact, moist, not frozen	None
1.8	GW Nf	Brown well-graded, sandy GRAVEL, frozen, poorly bonded	No visible segregation, negligible thin ice film on gravel sizes and within larger voids
3.7	GW Nbn	Brown, well-graded, sandy GRAVEL, frozen, well bonded	No visible segregation
5.4	ML Vs	Black, micaceous, sandy SILT, frozen	Stratified horizontal ice lenses averaging 4 inches in horizontal extent, hairline to $\frac{1}{4}$ inch in thickness, $\frac{1}{2}$ to $\frac{3}{4}$ inch spacing. Visible excess ice ~ 20±% of total volume. Ice lenses hard, clear, colorless.
7.7	ICE		Hard, slightly cloudy, colorless, few scattered inclusions of silty SAND
9.1		Dark brown PEAT, frozen, well bonded, high degree of saturation	~ 5% visible ice
10.5	MH Vr	Light brown SILT, frozen	Irregularly oriented ice lenses and layers $\frac{1}{4}$ to $\frac{3}{4}$ inch thick on random pattern grid approx. 3 to 4 inch spacing. Visible ice ~ 10±% of total volume. Ice moderately soft, porous, gray-white.
14.3		Bedrock. Laminated SHALE Top few feet weathered	1/16 inch thick ice lenses in fissures to 16.0 ft. None below
16.0			
20.6		Bottom of exploration	

* Surface elevation 963.2 ft

Fig. 12.4. Example of the use of the frozen soil classification system in typical exploration (Linell and Kaplar 1966. Adapted from *Permafrost: Proceedings of an International Conference,* p. 486, with the permission of the National Academy of Sciences, Washington, D.C.).

ment Station (1953). Recommendations concerning the relative desirability of soils of the various groups for residential site development, including roadways, foundations, and water storage and sewerage facilities, have been published by the U.S. Federal Housing Administration (1959).

The AASHTO soil classification system

The American Association of State Highway and Transportation Officials Classification System was introduced originally by the U.S. Bureau of Public Roads in 1928. Its primary use has been to classify soils according to their engineering characteristics and performance as subgrade materials under pavements. Familiarity with this system is widespread among highway agencies in North America, including organizations in both the United States and Canada. Classification of soils in accordance with this system may often be desirable or necessary if the results will be used by highway or airfield engineers experienced in the use of the AASHTO system.

The system has undergone many revisions since it was first introduced by the Bureau of Public Roads. The current classification procedure, AASHTO Designation M145-73, 'Recommended Practice for the Classification of Soils and Soil-Aggregate Mixture of Highway Construction Purposes' (American Association of State Highway and Transportation Officials 1978), uses sieve analysis, liquid limit, and plasticity index test data, as indicated in Tables 12.6(a) and (b). Classification into seven major groups is shown in Table 12.6(a). If more detailed classification is desired, further subdivision of the groups may be made as shown in Table 12.6(b). A soil is classified by comparing the appropriate test data from material passing the 75 mm sieve with the specification limits in Table 12.6(a) or (b), progressing from the left side of the table toward the right. The classification corresponds to the group heading at the top of the first column from the left in which the test data fit the specification. Figure 12.5 shows graphically the liquid limit and plasticity index specification ranges for the silt-clay materials.

Highly organic materials, including peat, may be classified in an A-8 group. Classification of these materials is based on visual inspection and is not dependent on sieve analysis, liquid limit, or plasticity index data.

A group index (GI) is used to provide an evaluation of soils within each group and is useful as an indicator of the relative quality of the soil material for use in earthwork structures, particularly embankments, subgrades, subbases, and bases. A group index of 0 indicates a good subgrade, and a group index of 20 or greater indicates a very poor subgrade material. The group index is given by the formula:

$$GI = (F\text{-}35)\,[0.2 + 0.005\,(LL\text{-}40)] + 0.01\,(F\text{-}15)\,(PI\text{-}10),$$

where F is the percentage passing 0.075 mm (No. 200) sieve, expressed as a whole number. This percentage is based only on the material passing the 75 mm sieve; LL is the liquid limit, and PI is the plasticity index.

TABLE 12.6(a).

Classification of soils and soil-aggregate mixtures under the AASHTO system

General classification	Granular materials (35% or less passing 0.075 mm)			Silt-clay materials (more than 35% passing 0.075 mm)			
Group classification	A-1	A-3[a]	A-2	A-4	A-5	A-6	A-7
Sieve analysis, percent passing:							
2.00 mm (No. 10)	....	....	....	....	....	....	....
0.425 mm (No. 40)	50 max.	51 min.	....	....	....	....	....
0.075 mm (No. 200)	25 max.	10 max.	35 max.	36 min.	36 min.	36 min.	36 min.
Characteristics of fraction passing 0.425 mm (No. 40)							
Liquid limit	....	....	[b]	40 max.	41 min.	40 max.	41 min.
Plasticity index	6 max.	N.P.	[b]	10 max.	10 max.	11 min.	11 min.
General rating as subgrade	Excellent to good			Fair to poor			

[a] The placing of A-3 before A-2 is necessary in the "left to right elimination process" and does not indicate superiority of A-3 over A-2.

[b] See Table 12-6(b) for values.

TABLE 12.6(b).

Classification of soils and soil-aggregate mixtures under the AASHTO system

General classification	Granular materials (35% or less passing 0.075 mm)							Silt-clay materials (more than 35% passing 0.075 mm)			
	A-1			A-2							A-7
Group classification	A-1-a	A-1-b	A-3	A-2-4	A-2-5	A-2-6	A-2-7	A-4	A-5	A-6	A-7-5, A-7-6
Sieve analysis, percent passing:											
2.00 mm (No. 10)	50 max.	....	....	....	....	....	....	....	....	....	....
0.425 mm (No. 40)	30 max.	50 max.	51 min.	....	....	....	....	....	....	....	....
0.075 mm (No. 200)	15 max.	25 max.	10 max.	35 max.	35 max.	35 max.	35 max.	36 min.	36 min.	36 min.	36 min.
Characteristics of fraction passing 0.425 mm (No. 40)											
Liquid limit	....		....	40 max.	41 min.	40 max.	41 min.	40 max.	41 min.	40 max.	41 min.
Plasticity index	6 max.		N.P.	10 max.	10 max.	11 min.	11 min.	10 max.	10 max.	11 min.	11 min.[c]
Usual types of significant constituent materials	Stone fragments, gravel and sand		Fine sand	Silty or clayey gravel and sand				Silty soils		Clayey soils	
General rating as subgrade	Excellent to good							Fair to poor			

[c] Plasticity index of A-7-5 subgroup is equal to or less than LL minus 30. Plasticity index of A-7-6 subgroup is greater than LL minus 30 (see Fig. 12.5).

Source: AASHTO 1978, Designation M145-73.

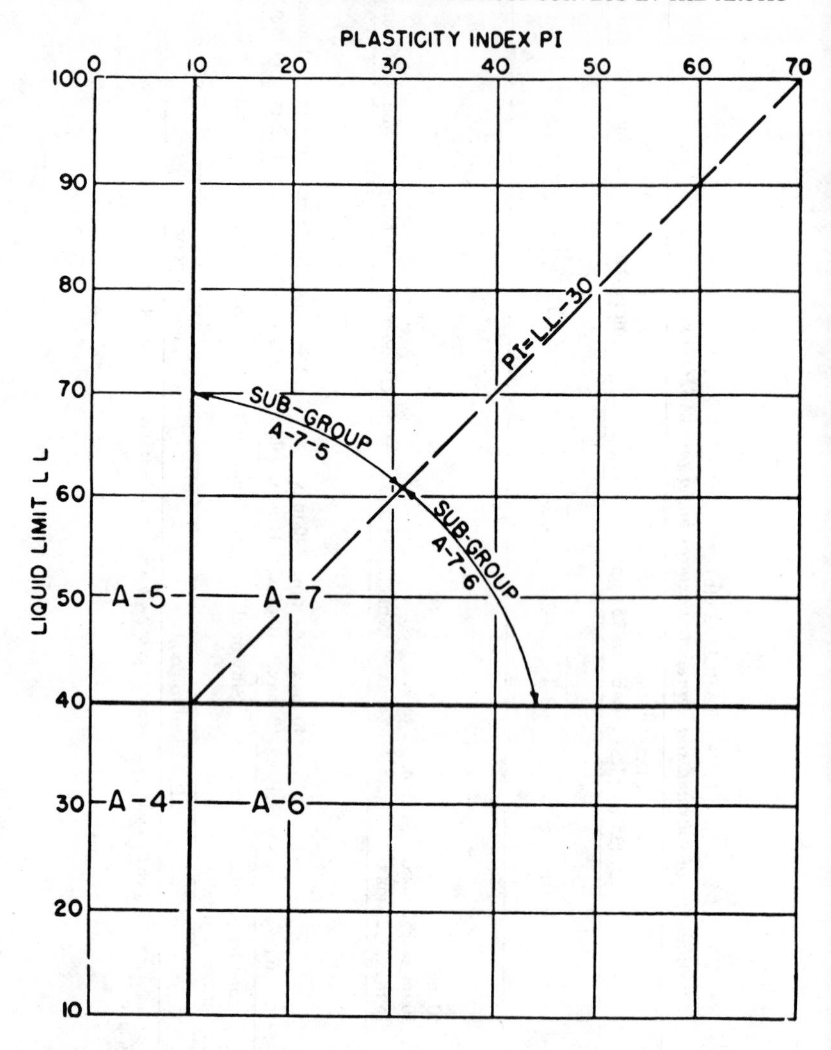

Fig. 12.5. Liquid limit and plasticity index ranges for subgrade groups A-4, A-5, A-6, and A-7. (AASHTO 1978, Designation M145-73.)

When the calculated group index is negative, the group index is reported as zero. The group index is reported to the nearest whole number and is shown in parentheses after the group or subgroup designation, as A-4(5). The group index should be given for each soil, even if the numerical value is zero. This will distinguish the classification from those made under the original Bureau of Public Roads system. The applicable AASHTO designation number should also be identified in the soils report, however, because of periodic changes which may be made under the AASHTO system.

The group index equation may be solved with the aid of the chart on Fig. 12.6 by determining the partial group index due to liquid limit and that due to plasticity index and adding them together. When determining the group index of A-2-6 and A-2-7 subgroups, only the PI portion of the formula (or of Fig. 12.6) is used.

General ratings of the materials as subgrade for engineering purposes are shown at the bottoms of Tables 12.6(a) and 12.6(b), and the usual types of significant constituent materials in the groups are shown in Table 12.6(b). More detailed explanation of the classification system and of characteristics of individual groups and subgroups may be obtained from AASHTO Designation M145-73.

As may be seen in Table 12.1, the Unified and AASHTO systems both use the 0.075 mm (200 mesh) sieve as the dividing point between silts and clays and the coarser materials. Under the AASHTO system, the term *silty* is applied to fine material having a plasticity index of 10 or smaller, and the term *clayey* is applied to fine material having a plasticity index of 11 or greater, whereas the Unified System employs the A-line system for distinguishing between fine-grained soils, as previously described. The two systems agree on the size range for fine sand, but differ again on the size classifications of other types of coarse materials. Thus, these two systems agree only in part on these basic points. As shown on Table 12.1, other size classification systems employ still other subdivisions. Finally, the American Society for Testing Materials (ASTM), 1979, has published both a standard corresponding to the AASHTO System, ASTM Designation D3282-73 'Standard Recommended Practice for Classification of Soils and Soil-Aggregate Mixtures for Highway Construction Purposes,' and one based upon the Unified System, ASTM Designation D2487-69 'Standard Method for Classification of Soils for Engineering Purposes.'

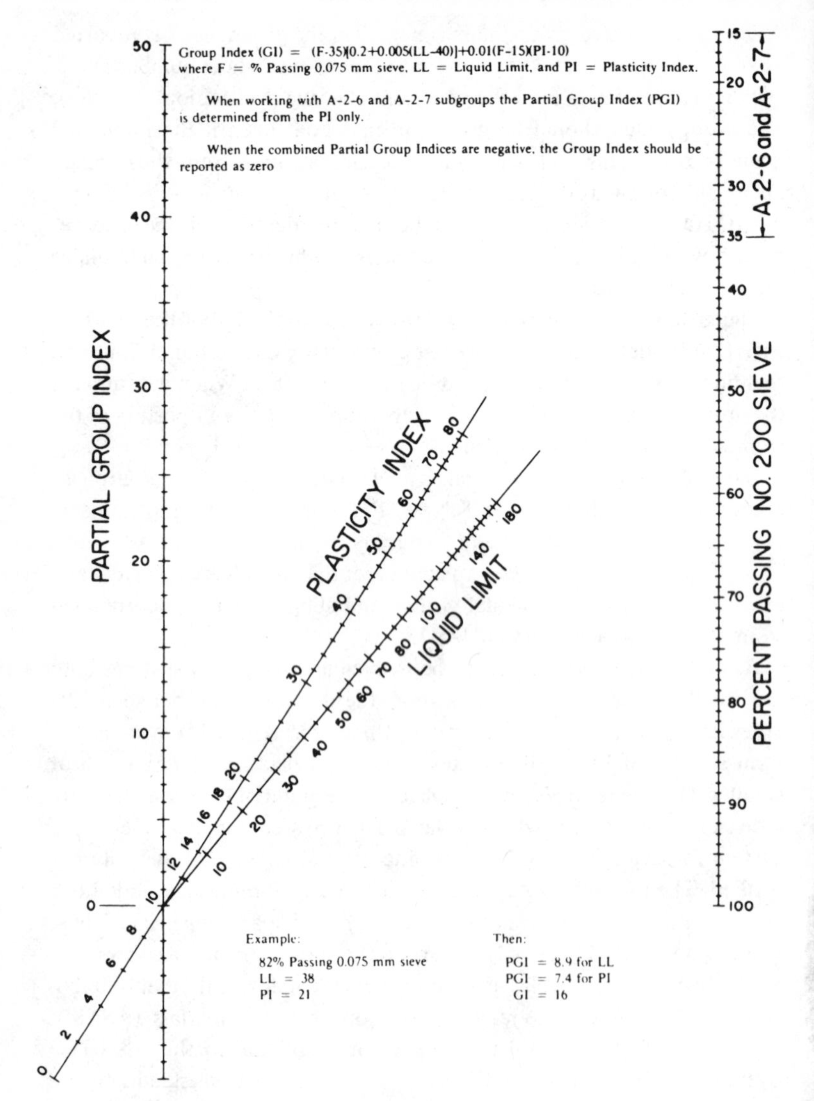

Fig. 12.6. Group-index chart (AASHTO 1978, Designation M145-73).

Other classification systems

In soil surveys for engineering projects, and during engineering design, an opportunity may sometimes arise to take advantage of existing data in which the materials are identified under nonengineering soil classification systems. In new surveys which may be used for multiple purposes, it may be desirable to classify materials under multiple systems, including nonengineering ones. Unfortunately, difficulties are encountered in attempting to translate between the present engineering classifications and others, like the pedological classification systems or, in some cases, older-edition engineering systems. Handy and Fenton (1977) have stated that it is unlikely that a precise translation between textural and engineering classifications will ever be made because their purposes differ. The engineering systems attempt to reflect engineering behavioral characteristics. The pedological systems, by contrast, have historically been strongly descriptive. The engineering classification systems are nongenetic, whereas the agricultural soil series system has been genetically based. The textural pedological systems are based solely on grain size, but the engineering classifications evaluate the fines on the basis of plasticity. The new *Soil Taxonomy* approach intends to reduce these differences by placing reliance on hard data, but most currently available data are likely to be expressed under the older systems. Because size limits for soil fractions and textural class boundaries may differ among both current and past systems, the survey team must be careful to ascertain the size range definitions for gravel, sand, and fines which actually applied for classifications reported in older data or under other systems.

One method that can be employed for making use of prior pedological soil survey information is to perform laboratory engineering classification tests on specimens selected to be representative of the pertinent pedological classifications. Table 12.7, which is taken from the FHA report 'Engineering Soil Classification for Residential Developments' (1959), shows an example of this approach as applied to temperate zone soils. The data contained in this report were obtained by the Division of Physical Research, Bureau of Public Roads, U.S. Department of Commerce, by testing soil samples collected by Soil Survey, Soil Conservation Service, U.S. Department of Agriculture. This new information can, if desired, then be used to prepare engineering soil maps from the soil survey information or it may simply be added to the existing maps.

It is anticipated that such engineering-related information will be included routinely in soil survey reports issued in the future. Nevertheless,

TABLE 12.7.

Example of engineering test data for soils sampled by Soil Conservation Service and tested by Bureau of Public Roads

Soil name	Location (county and state)	Parent material	Position from ground surface		Moisture-density		Liquid limit
			Depth	Horizon[1]	Maximum dry density	Optimum moisture	
			Inches		Lb./cu. ft.	Percent	
Silerton silt loam	Henderson, Tenn.	Loess over Coastal Plain	0–7		104	16	26
			7–24		104	20	45
			24+		110	16	43
Sinai silty clay loam	Brookings, S.D.	Lacustrine silty clay on Cary drift plain	0–8	A	92	23	48
			8–17	B	98	21	54
			17–29	B_{ca}	105	19	49
			29–60	C_{ca} & C_g	101	22	54
Skyberg silt loam	Fillmore, Minn.	Glacial till	7–12	A	108	17	30
			18–27	B	119	12	25
			27+	C	122	11	25
	Fillmore, Minn.	Glacial till	5–12	A	111	15	27
			21–30	B	121	11	22
			30+	C	121	12	25
Sodus gravelly fine sandy loam	Wayne, N.Y.	Glacial till	0–8	A	105	13	NP
			11–16	B	112	13	NP
			16–28	A_b	120	9	NP
			28–58	B_b	122	9	16
			58–70	C	124	9	16
	Wayne, N.Y.	Same	0–7	A	110	13	26
			8–21	B	120	10	17
			21–29	A_b	122	9	16
			29–45	B_b	124	9	17
			45–57	C	124	9	16
	Wayne, N.Y.	Same	0–5	A	106	15	27
			8–18	B	112	13	NP
			18–26	A_b	122	10	NP
			26–85	B_b	123	10	16
			85–121	C	120	9	15
Staser fine sandy loam	Blount, Tenn.	Alluvium	0–14		107	18	35
			14–36		116	14	26
Steinauer loam	Monona, Iowa	Glacial till	24–30		113	16	41
Suffield silt loam	Chittenden, Vt.	Lake deposit	0–6	A	93	23	40
			10–28	C	114	16	32
			28+	D	97	25	64
	Washington, Vt.	Glacial lake deposit	0–3	A	102	20	37
			7–21	B	104	20	32
			29+	C	99	22	NP
	Strafford, N.H.	Same	0–9	A	83	30	50
			12–17	B	99	22	32
			22–30	C_g	107	20	40

Source: Adapted from U.S. Federal Housing Administration 1959.

[1] Subscripts used with soil horizon symbols indicate special conditions as follows:

b—buried soil horizon

ca—a layer of accumulated calcium carbonate

g—a layer of reduction characterized by the presence of ferrous iron and neutral gray colors produced by a p involving saturation of the soil with water for long periods in the presence of organic matter.

[2] The American Association of State Highway Officials (AASHO) was the predecessor of the American Association o Highway and Transportation Officials (AASHTO). The AASHTO classifications were made in accordance with the no perseded AASHO Designation M145–49.

Mechanical analysis								Classification	
Percentage passing sieve					Percentage smaller than				
3-in.	No. 4 (4.7 mm)	No. 10 (2.0 mm)	No. 40 (.42 mm)	No. 200 (.075 mm)	.05 mm	.005 mm	.002 mm	A.A.S.H.O.[2]	Unified
		100	99	97	89	23	17	A-4 (8)	ML-CL
			100	99	98	41	36	A-7-6 (13)	ML-CL
100	95	89	84	80	71	36	30	A-7-6 (13)	CL
		100	98	89	81	42	34	A-7-6 (13)	ML-CL
			100	93	87	46	38	A-7-6 (18)	CH
			100	98	93	53	40	A-7-6 (16)	CL
			100	97	94	55	42	A-7-6 (19)	CH
		100	96	84	82	34	26	A-4 (8)	CL
	100	99	95	52	50	31	26	A-4 (3)	CL
		100	97	52	50	33	26	A-6 (4)	CL
100	90	99	91	70	38	32	23	A-4 (7)	
100	90	94	79	38	34	19	15	A-4 (1)	CL
100	90	98	84	47	43	22	17	A-4 (2)	SM-SC
100	70	77	74	49	35	8	5	A-4 (3)	SM
100	82	79	74	45	35	12	7	A-4 (2)	SM
100	85	83	77	49	34	11	6	A-4 (3)	SM
100	85	82	76	48	33	11	7	A-4 (3)	SM
100	90	87	81	51	38	12	9	A-4 (3)	ML
100	65	63	58	36	28	8	5	A-4 (0)	SM-SC
100	83	79	72	41	32	9	7	A-4 (1)	SM
100	84	81	73	44	33	11	7	A-4 (2)	SM
100	79	76	68	46	34	12	8	A-4 (2)	SM
100	91	89	84	56	41	15	8	A-4 (4)	ML
100	90	87	82	55	41	14	9	A-4 (4)	ML
100	83	80	75	48	36	12	8	A-4 (3)	SM
100	87	85	81	50	36	13	8	A-4 (3)	SM
100	90	88	84	56	42	11	7	A-4 (4)	ML
100	90	87	81	47	35	10	6	A-4 (2)	SM
		100	94	57	55	27	21	A-4 (4)	ML-CL
		100	89	42	42	17	14	A-4 (1)	SM-SC
100	99	98	92	72	69	39	32	A-7-6 (13)	CL
		100	99	94	83	26	14	A-4 (8)	ML
			100	97	89	39	28	A-6 (9)	CL
			100	99	97	85	73	A-7-6 (20)	CH
		100	97	91	88	40	26	A-6 (8)	ML-CL
			100	99	96	33	19	A-4 (8)	ML-CL
		100		100	98	30	13	A-4 (8)	ML
		100	96	86	81	40	26	A-5 (8)	ML
		100	99	98	94	39	25	A-4 (8)	ML
		100	99	98	93	51	38	A-6 (11)	CL

the pedological reports will always have the limitation that they are concerned primarily with approximately the upper 1 m of the soil cover, whereas the geotechnical engineer knows from experience that the deeper materials, which may be of primary interest to him, may be quite different.

13. Specific engineering applications

To ensure that surveys will include all the required information, the survey staff should be aware of the kinds of data needed and the information requirements for a variety of specific engineering features. Engineering features in cold regions must be so designed and constructed that they will not be adversely affected by such factors as frost-heaving displacements and forces, thaw-weakening, permafrost degradation and thaw-settlement, thermal shrinkage and expansion, downslope movements, and adverse drainage conditions.

Site or route data, engineering policies, criteria, and other constraints, knowledge of the state of the art, and facility technical data and requirements are used as shown diagrammatically in Fig. 13.1 to develop designs for engineering projects in frost and permafrost areas, with feedback a continuing feature at all stages of the design process, as

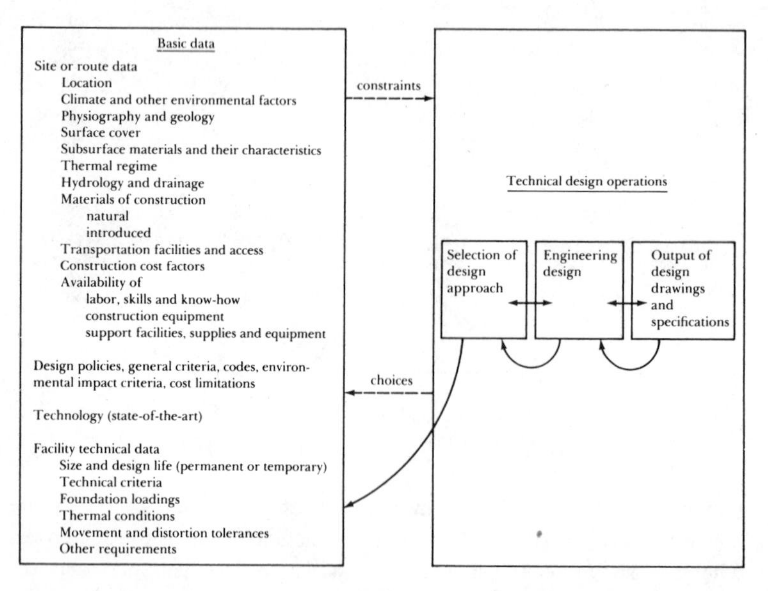

Fig. 13.1. Engineering design in areas of deep frost penetration and permafrost. (Modified from Linell, Lobacz, et al. 1980).

indicated by the arrows on the figure. The objective is to develop a final design that will best meet the project requirements at minimum cost.

Data requirements

Engineering projects involving soils in the arctic and subarctic may include construction features such as the following:

Foundations for structures.
Highways, airfields, and railroads.
Water-related facilities and construction.
Walls and retaining structures.
Utility lines and pipelines.
Excavations and embankments.
Shafts and tunnels.

It is impossible to list the data requirements for every individual type of project and situation. For example, some projects may be located in virtually unexplored areas for which little or no data are available. Others may be located in areas where the general conditions are already well known and only verification for the specific site is needed. Again, the data requirements for simple or temporary construction features may be very different than for complex and expensive types of permanent construction. Further, different types of engineering works may affect permafrost in different ways and require different degrees of investigation.

Table 13.1 presents a general check list of types of data, either observed or estimated, that may be needed for soil mechanics design on arctic and subarctic engineering projects. As appropriate, data from this list may be presented on soil profiles, in analyses of design, and in contract documents.

The items listed in the first section of Table 13.1 include the information essential for almost any engineering analysis or design study. Even some of these data may be unnecessary under certain conditions, however. For example, the position of bedrock may be of no consequence in the design of a small structure if the overburden is known to be very deep.

General considerations for design and construction in arctic and subarctic areas

To illustrate the types of choices involved during design of arctic and subarctic engineering projects, Table 13.2 shows, as an example, design

TABLE 13.1.
Data that may be required for soil mechanics design in arctic and subarctic areas

Most important data requirements
Surface freezing and thawing indexes, from air indexes and n factors, before construction and as modified by construction
Active layer thickness, before construction and as modified by construction
Position of permafrost table, before construction and as modified by construction
Positions of residual thaw zones, if any
Position of bedrock
Identification and classification of subsurface materials, including bedrock (includes gradations and Atterberg limits)
Densities, moisture or ice contents, and specific gravities
Ground temperature with depth
Position of water table, or artesian pressure level if any
Properties of borrow, backfill, concrete aggregate, and pile slurry materials
Thermal properties of soils (usually obtained from available charts)
Frost susceptibility of materials subject to seasonal freeze and thaw
Thaw consolidation and thaw settlement characteristics of permafrost
Permeability, unfrozen soil
Possible additional data requirements
Allowable bearing values, from unconfined compression creep tests
Sustainable tangential adfreeze bond strengths
Sustainable shear strength values
California bearing ratio values
Subgrade modulus values
Dynamic response characteristics
Field bearing test data, piles or footings
Field anchor test data
Swell potential, if any
Excavation, handling, and transport characteristics of soil and rock
Organic content
Presence of methane

alternatives for foundations for structures. For other features, such as utilities; highways, airfields, and railroads; excavations and embankments; and shafts and tunnels, the details of choices will be different, but the problem category divisions and basic solution assumptions will be the same.

Construction in areas of only seasonal frost

In discontinuous permafrost areas, permafrost may be absent under substantial areas of the terrain. In such locations, seasonal freeze-and-thaw effects may be expected to be severe. Engineering measures required to cope with the effects of deep, seasonal freezing and thawing will depend on the type of engineering feature involved and whether or not the subsurface materials are frost-susceptible. Adverse effects can be prevented or limited by measures such as the following, or combinations thereof:

Supporting foundations or placing utilities and pipelines below the maximum depth of frost penetration.

Restricting soils within the depth of frost penetration to non-frost-susceptible types.

Using thermal insulation to reduce or prevent frost penetration into frost-susceptible materials or to prevent freezing or excessive cooling of fluids in pipes.

Heating and maintaining circulation of fluids in pipes.

Using surcharge loading to reduce frost heave.

Reducing or preventing effects of thaw-weakening by placing sufficient cover of nonweakening materials over frostsusceptible soils.

Reducing or preventing adhesion of frost-heaving soil to piles or other structural elements.

Especially difficult design problems may arise when a building foundation is situated partially on permafrost, partially on nonpermafrost.

Construction on permafrost that will not be adversely affected by thaw

Construction on permafrost can usually be greatly simplified by selecting locations where the foundation materials are clean, granular, and non-frost-susceptible and free of excess ice. Design of such features as foundations and pavements can then frequently be essentially identical with temperate zone practice, even though the underlying ground is permanently frozen. Frost heave, thaw-settlement, and thaw-weakening problems may be absent or negligible in such materials. It is one of the tasks of preliminary and detailed site investigations to locate such desirable areas, if present, and to verify that such desirable conditions do exist, in relation to the proposed construction, if such locations are selected.

The Corps of Engineers has shown, in its construction in Central Alaska, that such favorable conditions do exist and that measures to preserve permafrost can then be unnecessary. Figure 13.2 indicates, for example, that only very minor settlement accompanied thaw progression under a three-story, reinforced concrete building at Fort Wainwright (formerly Ladd AFB), Fairbanks, Alaska. No adverse structural effects could be detected after final disappearance of the permafrost. Sometimes

TABLE 13.2.

Design alternatives for foundations

Foundation conditions adversely affected by freeze or thaw	Foundation conditions not adversely affected by freeze or thaw	Foundation conditions adversely affected by thaw
SEASONAL FROST	SEASONAL FROST / PERMAFROST	PERMAFROST
[Usually fine-grained soils, or rock containing mud seams]	[Clean, granular soils or rock w/o ground ice]	[Usually fine-grained soils or rock containing excess ice]
Use conventional foundations supported below annual frost zone and protected against uplift by adfreeze grip and against frost overturning or sliding forces or Support structure on compacted non-frost susceptible fill adequate to control frost effects or Use insulation, heat, or surcharge loading	Use Normal Temperate Zone Approach	

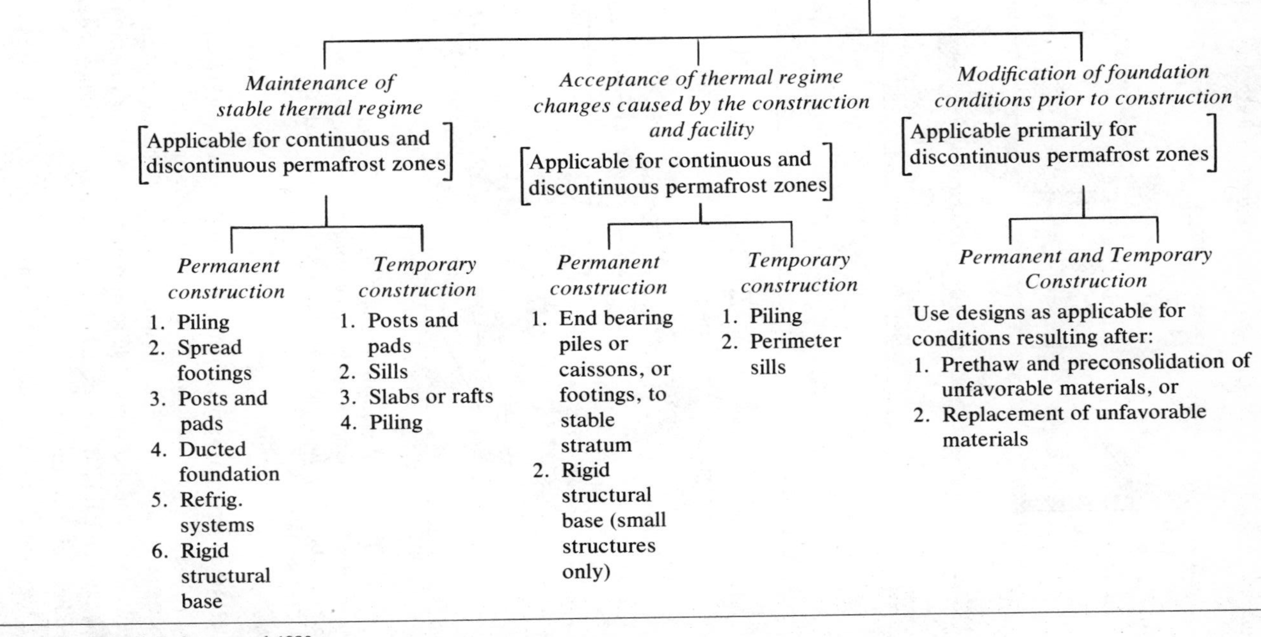

Modified from Linell, Lobacz, et al 1980.

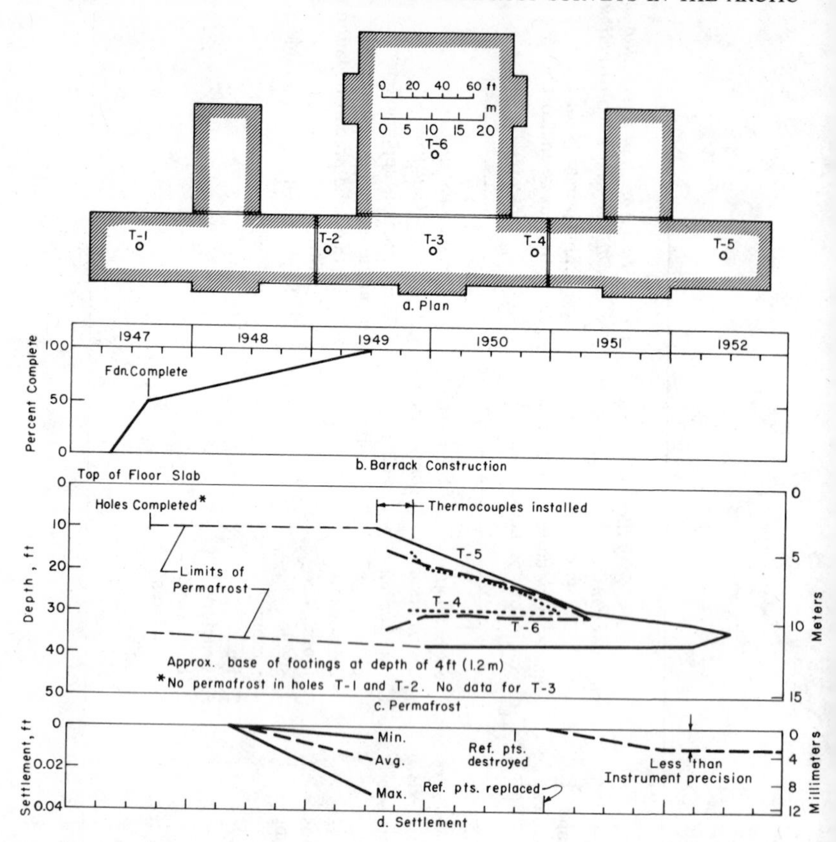

Fig. 13.2. Thawing of permafrost under three-story, reinforced concrete building, Fairbanks, Alaska. (Source: Thaw Penetration and Subsidence, 500-Man Barracks (Buildings 1001 and 1004), Ladd AFB, Alaska. CRREL Internal Report IR 12, 1966.)

such granular deposits may be loose enough, particularly in the upper 10 m, such that dynamic loading, as from a vibrating power plant foundation, may cause unacceptable densification of the supporting soil after thaw. Assessment of a possible requirement for preconstruction thawing and consolidation of the foundation in such a case may require very detailed soil investigations.

There have also been numerous reports of massive deposits of ground ice within clean, granular soil formations, the frequency of such occurrences apparently being greater the colder the climate of the location. It is not safe to depend on surface appearances or relatively shallow explorations, as ice may be present in underlying strata.

Ice-free bedrock can also provide excellent foundation conditions in permafrost areas. Again, however, an ice-free condition cannot be automatically assumed in the absence of frozen cores which can reveal any ice strata present and their thicknesses. Substantial construction problems have been encountered in both Alaska and Greenland from the too easy assumption that bedrock is ice-free.

Construction on permafrost that will be adversely affected by thaw

For permanently frozen soil or bedrock containing significant amounts of excess ice, three design alternatives for constructed facilities are available.

Maintenance of stable thermal regime. Under this approach, progressive thawing of permafrost and settlement are prevented, and permafrost temperatures are not allowed to warm above levels required for structural stability. This approach is by far the most commonly used and most acceptable method. It is applicable for both continuous and discontinuous permafrost zones.

Acceptance of thermal regime changes. Under this approach, permafrost is allowed to degrade, and any adverse consequences are accepted. Because the result, when permafrost contains excess ice, may be structural damage, poor serviceability, and high maintenance costs, opportunities for using this approach are limited except for very temporary construction and sometimes for slopes of cuts. It is applicable for both continuous and discontinuous permafrost zones.

Modification of foundation conditions prior to construction. Under this approach, the subsurface conditions or thermal regime are changed in advance of construction so that there will be no adverse effects after construction. This includes the alternatives of (i) removing and replacing thaw-unstable foundation materials and (ii) thawing and consolidating permafrost in place to a stable condition. Favorable conditions for employing this approach are somewhat rare. It is applicable mostly for discontinuous or sporadic areas of permafrost.

Each of these approaches requires detailed information on surface, subsurface, and climatic conditions for the performance of the construction to be predictable to an acceptable degree of accuracy.

Granular mats

Clean, granular non-frost-susceptible materials are useful for roads, working pads, and other uses during construction. When used as cover layers on foundations, they serve to modify freeze and thaw penetration into underlying frost-susceptible materials, as well as to apply surcharge loading on these materials, thereby reducing seasonal frost heave and thaw-weakening effects. Sources of borrow material for mats must be located and the material tested to determine suitability. Allowable bearing values on gravel materials may be based on the unfrozen condition, provided the thickness of granular material between the foundation and underlying non-frost-susceptible soil is sufficient to reduce stresses on the latter materials during thaw to tolerable levels.

Footings, rafts, and piers

A *footing* is an enlargement of a column or wall to distribute concentrated loads over a sufficient area so that allowable pressure on the soil will not be exceeded. A *raft,* or *mat, foundation* is a combined footing which covers the entire area beneath the structure and supports the walls and structural columns. A *pier* is a prismatic or cylindrical column that serves, like a pile or pile cluster, to transfer loads to a suitable bearing stratum at depth. For satisfactory performance on frost-susceptible or thaw-unstable materials, these types of structural foundations are supported, in seasonal frost areas, below the maximum depth of seasonal frost penetration and, in permafrost areas, on frozen soil below the level of any future degradation. Bearing pressures on permafrost must be limited to values which will not produce unacceptable, permanent, downward displacement of the supporting material over the life of the structure. The foundations must be sufficiently anchored, embedded, loaded, or isolated from heave forces acting on them in the zone of annual freezing so that they will not be lifted by frost heave. Sometimes footing or raft foundations may be supported in or on top of a gravel pad placed on top of the active layer, provided the effects of the consequent seasonal heave and settlement are tolerable for the particular supported structure.

Data of the type shown in the first section of Table 13.1, together with supplementary information on allowable bearing values and sustainable tangential adfreeze bond strengths, will usually be sufficient to accomplish design. Empirical data on allowable bearing values and sustainable tangential adfreeze bond strengths may often suffice for smaller projects. Dynamic response information and special field bearing tests may be needed in some cases.

Piling

A pile is essentially a very slender pier. Properly constructed pile foundations can isolate the structure from the seasonal heave and subsidence of the active layer and from at least limited degradation of the permafrost, can be built with minimum disturbance to the thermal regime, and can be easily adapted to provision of foundation ventilation. Accurate ground temperature and soil information is very important for design and construction of pile foundations. Piles are most commonly installed in frozen ground by inserting them in bored holes, placing soil and water slurry in the annular space around the piles, and allowing the slurry to freeze back. The permafrost temperatures at the time of construction are critical in determining the rate of slurry freeze-back, and the permafrost temperatures at their warmest extreme between the end of summer and early winter are critical in determining the sustainable tangential adfreeze shear strength values that can be relied on to provide the required bearing capacity and resistance to frost heave for friction type piles. In marginal permafrost areas, permafrost temperatures may be low enough to provide reasonable rates of freeze-back of slurried piles only during a few months of late winter and spring, unless mechanical refrigeration is employed. Under some conditions steel piles can be installed in frozen fine-grained soils by driving. Driving of piles into permafrost is likely to be impossible at ground temperatures below about 25°F (0°C). The temperature and other characteristics of frozen soil surrounding piles are of much less importance for piles designed to carry their loads in point bearing than for friction type piles, except that adequate lateral support must be provided for the piling. The strata supporting point bearing piles must be very competent, however; generally, soil or rock containing excess ice will not be suitable because of the rapid plastic deformation that would occur under the highly concentrated stresses imposed by a point bearing pile.

Of the data requirements shown on Table 13.1, the position of the permafrost table, the ground temperatures with depth, and the sustainable tangential adfreeze bond strengths are of key importance. Frost susceptibility of the active layer soils is also a significant consideration for pile foundations, as the piles must have sufficient depth of embedment in the underlying frozen materials to resist frost-heave forces that may act on the piles. Air temperatures and wind velocities are also important in calculating the effects of thermal piles, which are designed to remove heat from the ground and dissipate it aboveground, either automatically or by forced internal circulation.

When field pile load tests are performed for design purposes, special procedures must be used because of the characteristic creep or slow plastic deformation of ice-saturated frozen soil. Conventional temperate zone procedures are not satisfactory. A pile deformation rate of 1/1000 in. (0.0254 mm) per day continued over 20 years would produce 7.3 in. (0.185 m) settlement of the supported structural element. Necessary modifications of conventional procedures are detailed in U.S. Army/ U.S. Air Force (forthcoming) Technical Manual TM5-852-4.

Anchorages

Ground anchors, as opposed to footings, piers, and bearing piles, are designed to withstand tensile forces such as from the cables of guy-supported towers. They may consist of various types, such as deadman, plate, screw-in, helical, mass-gravity, or various patented anchors. Grouted-rod anchors may be used in ice-free bedrock if the rock temperatures are above 30°F (−1.1°C) and precautions are taken so that the grout will set and gain strength rapidly without freezing. Conventional piles may also be used, but it should be noted that applied tensile forces and frost-heave forces are then additive. As with footings, failure occurs in creep. Soil shear strength, tangential adfreeze bond strength, and dry unit weight are important soil properties, depending on the type of anchor and depth of placement below ground surface. The mode of failure is different for deep plate anchors than shallow plate anchors. The strength values, of course, are functions of the ground temperature and are very important, except for surface mass-gravity anchors.

Highways, airfields, and railroads

These types of facilities must, by the nature of their functions, be supported principally on the ground surface, whether at grade, in cut sections, or on embankments. They are thus directly exposed to full seasonal temperature variations and freeze-and-thaw effects. Design objectives are (i) to provide adequate load-supporting capacity for the anticipated traffic under seasonal fluctuations of support, (ii) to ensure an adequately smooth operating surface under effects of freeze and thaw, and (iii) to control permafrost degradation and thaw settlement. Design techniques for pavements have been described by Linell et al. (1963).

Soil studies for these types of facilities may range from route and location studies for entirely new projects to route or location modifications of existing facilities or investigation of causes for soil problems in operating facilities.

These types of facilities commonly extend over long distances or over large ground areas with usually relatively shallow depth involvement except at special locations such as bridges or deep cuts. The earth and rock materials and the permafrost and drainage conditions that will be encountered are major considerations in route selection. The types of data shown in the first section of Table 13.1, supplemented as necessary by special design tests, such as modulus of soil reaction or California bearing ratio tests will serve for design purposes.

Water-related facilities and construction

Especially difficult problems may arise when construction sites are located in or close to bodies of surface water, such as rivers, lakes, channels, or reservoirs. These may include such facilities as bridges, culverts, wharves, dams, and water-supply structures. The occurrence, temperature, and continuity of permafrost is greatly affected by the presence of surface water deep enough to restrict or prevent the penetration of freezing temperatures into the underlying ground in winter. In delta or meander areas of rivers such conditions are very common. Under very large water bodies, permafrost may be absent except close to the shore. Where permafrost exists it may then be very warm and susceptible to thawing and thaw-settlement. The effective tangential, adfreeze bond strength may be weak and unreliable, making pile foundations susceptible to frost heave or settlement. Figure 13.3 shows, for example, a typical bridge foundation on the Alaska Railroad where, as reported by Péwé and Paige (1963), continual heaving of the foundation piles has required installation of replacement pile bents; the depression of the permafrost table under the stream and the inadequate or nonexistent depth of embedment in permafrost of some of the piles are clearly evident. Variations of water levels, snow cover, and winter icing accumulations may contribute to variations from season to season in subsurface thermal conditions. Groundwater movements adjacent to water bodies and seepage through embankments and foundations of dams may contribute to

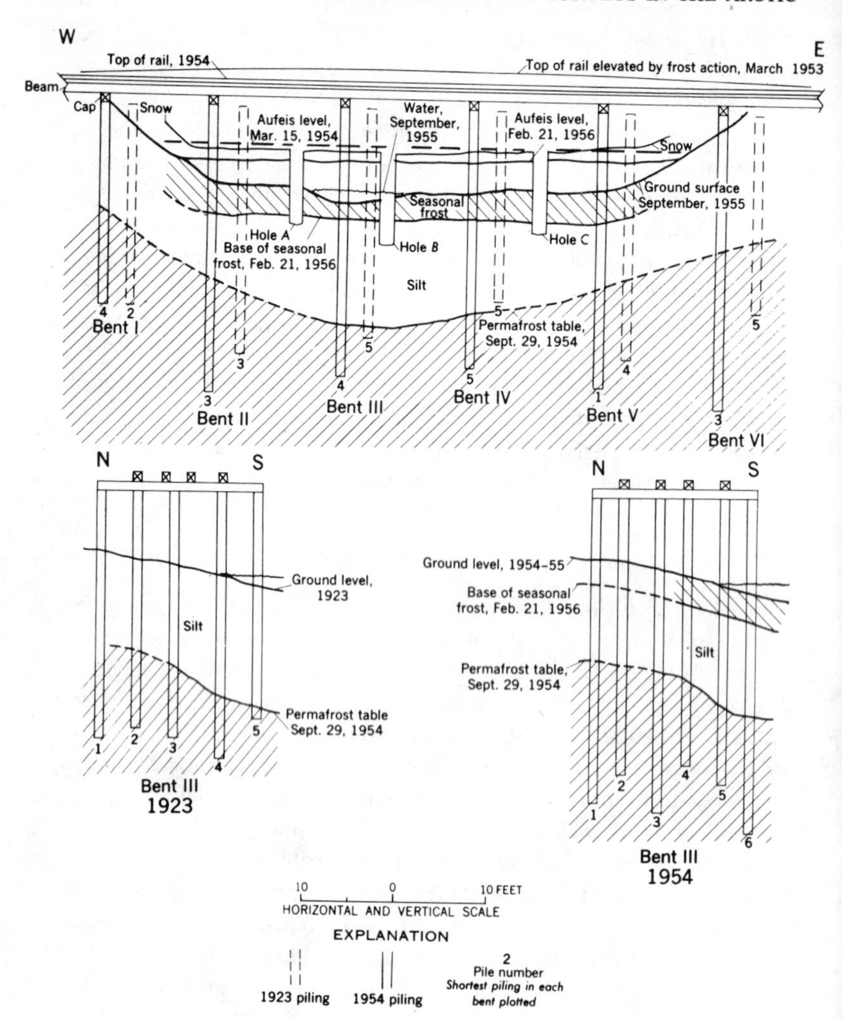

Fig. 13.3. Geologic section at a railroad bridge in Goldstream Valley near Fairbanks, Alaska (Péwé and Paige 1963).

complex thermal regime patterns and to permafrost degradation. Where permafrost is found offshore in northern coastal waters, special thermal conditions exist; although the sea bottom is separated from contact with air temperatures by overlying seawater, the water temperature may drop as low as -1.9°C without freezing because of the salinity of the water.

Because of the complexity and variability of such permafrost conditions, very careful site investigations are required when construction is to be carried out near the edges of or in bodies of surface water. Determination of the position of the permafrost table, the positions of any residual thaw zones, and the temperature of the permafrost are especially important, together with estimation of possible variations of these conditions during the life of the facility. The presence of springs or other evidences of subsurface water movement which may affect the thermal stability of the foundation should be noted.

Walls and retaining structures

Bridge abutments, retaining walls, bulkheads, and similar structures must be designed to control frost heave, lateral frost thrust, and settlement if the requisite soil, moisture, and freezing or thawing conditions are present.

As indicated in Fig. 8.5(b), frost-heave forces develop in a direction parallel to the direction of heat flow. Such forces developed below the base of a wall will tend to lift it, and forces developed behind the vertical face of a wall will tend to overturn it, shove it laterally, or even fracture it. Progressive, small, annual increments of tilting can accumulate to large, permanent deformations. Some techniques for coping with such effects are illustrated in Fig. 13.4.

Data needed for design are the same as for footings, rafts, and piers, except that dynamic response characteristics are unlikely to be needed.

Utility lines and pipelines

Utility lines serving communities (water, sewer, gas, electric, steam, fuel, and communications) and oil and gas pipelines are much more difficult and expensive to construct and maintain in the arctic and subarctic than in the temperate zones.

In order to place utility lines and pipelines underground in the arctic or subarctic, risk of damage from frost heave, thaw-settlement, overstressing where the lines cross ground-shrinkage cracks, and loss of thermal properties in insulating materials by absorption of moisture must be controlled. When lines are placed underground, the depth of burial in areas of permafrost or very deep seasonal freezing may be made nominal or minimal wherever possible in order to simplify access for maintenance and repair, but possibility of frost heave, thermal contraction, and

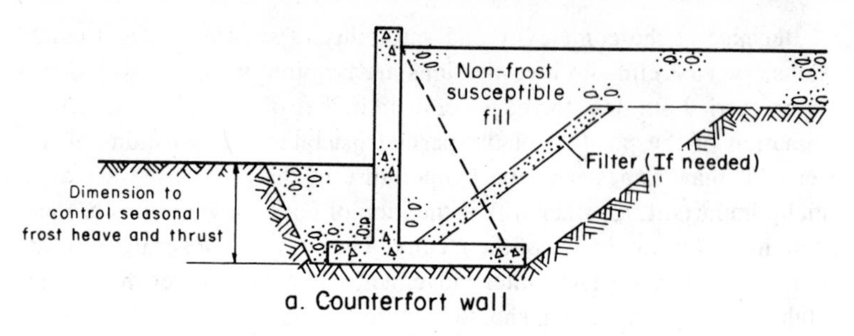

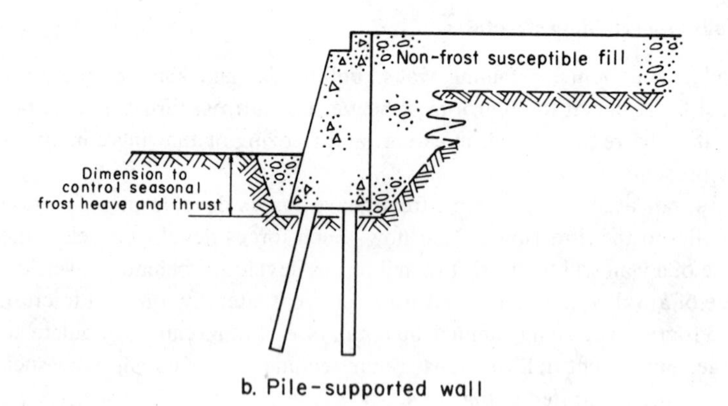

Fig. 13.4. Wall treatments (Linell, Lobacz, et al. 1980).

freeze-up damage must be considered. Where conditions are unfavorable for underground emplacement or are more expensive, water, sewer, and steam lines are commonly carried aboveground within insulated utilidors supported on cribbing, pads of granular soil, or piles; petroleum pipelines are also commonly supported on piles. Overhead electric and telephone lines are carried on poles anchored against heave by an adequate depth of embedment in permafrost or other means or supported on top of the active layer with the aid of such techniques as crossed-timber bases, rock-filled timber cribs, or erection of the poles in groups of three to form tripods. Poles conventionally installed may be frost-heaved out of frost-susceptible soils in only a few years.

The soil information needed for design will consist of pertinent elements of the data listed in Table 13.1 and will vary with the design alternatives selected for consideration. Soil data will usually be required to a relatively shallow depth over substantial horizontal distances and may involve many variations in terrain conditions and construction methods.

Excavations and embankments

Excavations may be needed for many different purposes, such as for emplacement of utilities underground, achieving desired grades on roads, highways, and airfields, preparing building sites, installing seepage control measures in dams, obtaining borrow materials, or excavating and removing frost-susceptible or ice-filled soils overlying more suitable foundation materials in order to simplify design and construction problems. The difficulty of excavation will vary widely with such factors as the types of materials involved, whether they are frozen or not, their ice contents and temperatures, and the presence of groundwater.

Soils encountered in permafrost areas may range from coarse bouldery soils through gravels, sands, silts, and clays to organic soils. At one extreme, coarse, bouldery soils at temperatures well below freezing may behave in excavation and tunneling as if they were granite, requiring systematic drilling and blasting. At the other extreme, clean gravel at low moisture contents may be excavated readily, even at low temperatures, and fat clays at temperatures warm enough that they contain little or no ice and soils with saline pore water may show little or no change from their behavior at temperatures just above freezing. In general, however, excavation in frozen materials, including moisture-bearing rock, can be much more difficult and expensive than when the same materials are unfrozen. Excavated frozen materials are often difficult to handle at below-freezing temperatures because of their tendency to freeze onto the surfaces of excavating, handling, transporting and processing equipment, and fine-grained soils may be unsuitable for use in compacted fills or embankments after thawing if substantial quantities of excess ice are present, which will make the thawed material too wet to be workable. Excavation into permafrost containing excess ice is avoided whenever possible because of these problems and because of the difficulty of avoiding the effects of degradation in the newly exposed permafrost.

It is not feasible to construct stable fills, backfills, or embankments for foundations, roads, earth dams, or other projects using frozen materials with substantial ice contents. Therefore, embankments are usually constructed with thawed active-layer materials occurring at acceptable moisture contents, during the summer, or of very low moisture content gravels or broken rock, which can be satisfactorily placed at below-freezing temperatures in other seasons. These are often stockpiled during the summer months to provide drainage of excess moisture in order to assist their use in colder months. Sometimes soils may be treated with salt solutions to prevent them from freezing during placement at below-freezing temperatures. Because active-layer materials thaw only gradually in summer and to limited depths, borrow areas on permafrost often have to be of large horizontal extent. In seacoast areas, gravel may be available in coastline deposits, but possible adverse environmental effects caused by removing such materials must be avoided.

The information required for design decisions is primarily that listed in the first section of Table 13.1. In some cases, field experiments to determine ripability (capacity to be loosened by tractor-drawn "ripping" equipment) or other excavation or emplacement characteristics may be appropriate. It is important to determine the moisture contents and compaction characteristics of materials which may be used in compacted fills. In remote areas, the types of compaction equipment which are available or can be brought to the site without undue difficulty may also need to be determined, as this may limit the degree of compaction that can be feasibly obtained.

Shafts and tunnels

Shafts and tunnels may be constructed in frozen ground for mining, refrigerated storage, or other purposes. Many of the problems connected with such work are similar to those described above for excavation, although the temperature and creep behavior of the materials are of special importance. Experience shows that tunnels in frozen materials may close progressively under the pressure from overlying materials, and the rate of closure increases with increase in soil temperature. Special measures to lower the temperatures in the walls of the tunnel or shaft may be necessary if the rates of closure are too rapid. On the other hand, some frozen soils, such as well-graded glacial moraine deposits at less than 100 percent ice saturation, may not exhibit any significant creep behavior.

Exploratory borings in advance of such construction should provide basic data which will permit anticipation of the types of materials that will be encountered, their probable behavior during the construction operations, and the tunnel support or temperature control measures that may be necessary after completion. Laboratory unconfined compression creep tests on core specimens of the frozen materials may be appropriate for some materials, but for others, such as cobbley or bouldery soils, only actual underground excavation may provide definitive information.

14. Field organization

The importance of thorough and detailed advance planning and careful organization of field activities for the success of arctic and subarctic soil investigations cannot be overemphasized. Planning must be exact and accurate down to the smallest detail. Major controlling factors in the success of a survey are the degree to which the work has been pre-planned and the knowledge of the area which the field workers have acquired before entering the field investigations (Johnston 1966).

Personnel

In order to ensure that the greatest amount of accurate and detailed information will be obtained, the survey team should be very carefully chosen so that the combined expertise will cover the entire spectrum of talents required for carrying out a comprehensive, high-quality investigation. It is essential that each specialist be highly qualified in his own field of investigation. The key personnel should be intimately involved in the planning of the investigation, to ensure that every detail of the field work is fully thought out and that no details are overlooked. The makeup of the team may vary with the scope and specific objectives of the investigation. If the objective is solely site selection and development of information needed for design of an engineering project, the team for an important project may typically include (a) an engineer familiar with the immediate and ultimate uses of the installation to act as coordinator, (b) a civil engineer (soil mechanics), (c) a civil engineer (hydrology), (d) a geologist, and (e) a forest ecologist. All should be familiar with permafrost. For a purely scientific study it will be unnecessary to include any engineering personnel in the team. The types of professional personnel then will be determined by the objective of the study, but a well-balanced team may typically include a pedologist, a geologist, a forest ecologist, and a hydrologist. If the investigation has both engineering and scientific purposes the team should be composed of a combination of such personnel. Depending on the scope of required investigations, its organization, and the location, services may also be needed of (i) a drilling and test-pit foreman, (ii) field laboratory technicians, (iii) equipment operators, and (iv) miscellaneous support personnel, such as local laborers, a cook, and possibly a guide. If the soil investigation is part of a comprehensive site-development study, at least one survey chief of party and a survey crew will probably need to be available for topographic and profile tie-ins. Especially during the reconnaissance stages of the study, the services of experts in aerial photography interpretation and remote sensing techniques may be needed. On the other hand, for projects of very limited scope the field work may have to be accomplished by as few as two or three men who must then combine in themselves all required functions, possibly with only aerial photographs or very rudimentary maps to provide orientation. Field crews of whatever size should have the ability to identify landforms on the ground and in aerial photographs in order to aid in their interpretation.

It is of course desirable that scientific personnel already have expertise in arctic and subarctic soil surveys and that engineering personnel be versed in engineering problems peculiar to arctic and subarctic regions. If there are any deficiencies among the selected team members in this regard, training should be provided. Any member who can not perform with full effectiveness not only fails to pull his own weight but may hinder the work of others by forcing them to make up for his deficiencies and by uselessly burdening sometimes limited support provisions. Training in aerial photography interpretation as applied to arctic and subarctic terrain may be particularly helpful for personnel not familiar with these techniques.

In selecting members of field teams who will be working and living very closely together for extended periods, their capacities to work harmoniously together under stressful conditions should be taken into account. Health requirements are discussed later.

Scheduling and access

In arctic and subarctic regions it is of utmost importance to schedule the various phases of soil investigations, site selection, and construction to conform with favorable, seasonal ground, weather, and access conditions and availability of support. Some operations cannot be conducted during certain seasons or can be done in such seasons only with excessive expenditure of time, effort, and funds. Summer is not the most favorable season for all activities or all observations, although terrain reconnaissance in the arctic and subarctic is carried out most effectively in the summer when the ground is snow-free. Aerial photographs and visual reconnaissance flights should not be attempted in the winter because of poor light conditions and because the surface cover of drifted snow and ice may obscure ground details. Flights made just previous to and during the breakup period, when icing and flooding conditions are at their worst, will give information on locations where winter icing formations occur, areas subject to seasonal flooding, and river areas subject to erosion by water and moving ice. Aerial photography reconnaissance during the spring snow-melting period may also be helpful in selecting locations for roads and facilities. As the spring snow-melting progresses and the general terrain begins to become free of snow, the areas of deeper snow accumulation tend to remain snow-covered longer. By identifying such areas, it may be possible to minimize winter snow control for new roads and facilities by avoiding their placement in these areas or by orienting them so as to be parallel, so far as possible, rather than at right angles to the prevailing directions of winter snow drifting.

Access to the area of investigation may be a major controlling factor in scheduling the various phases of the field studies. Movement of vehicles and equipment over trailless terrain is virtually impossible during the spring breakup, and may cause severe and long-lasting environmental damage to sensitive terrain at any time during the summer. In Alaska, operations in summer on the natural tundra surface are prohibited. This prohibition has led to the adoption of such measures as moving exploration rigs to drilling locations by helicopter, even to attaching a drill rig to the undercarriage of a helicopter, so that borings can be made in summer without surface travel, if schedules do not permit waiting for winter conditions. In Canada, land-use regulations must be complied with by anyone going anywhere on the lands north of 60°, and a specific land-use permit is required for every activity that involves going on the surface of the land at any time of the year (Linell and Johnston 1973). Other restrictions on entry into the proposed area of investigation may exist which will require various permits or authorizations to be obtained. Cross-country travel is most readily accomplished in most areas after the freeze-up period when water bodies have developed safe ice-cover thicknesses, the upper soil strata have refrozen, and a snow cover has accumulated over low vegetation. Extremely low temperatures may hinder mechanical operation of motorized equipment. Frozen lakes and rivers can provide excellent routes for surface transportation in winter, whether for tractor-drawn, double-end sleds and wanigans, over-snow personnel vehicles, or simply men on skis.

The thickness of floating ice required to support a given load is affected by many variables, such as the type, temperature, and quality of the ice, the effects of thermal stress in the ice sheet, the geometrical dimensions of the load, the spacing between loads, whether the load is moving or stationary, the velocity of travel of moving loads, the duration of stationary loading, the depth of water under moving loads, the number of load repetitions, and the presence of through cracks or a free ice edge close to the point of load application. Care must be taken to avoid breaking through at points of weakness or thinner ice resulting from springs, currents, shoreline seepage emergence, ice movements, extra snow depth, or other causes. Though obtaining an accurate picture of the ice thickness variations along a proposed route may take time, the effort can be very worthwhile. Thinner ice sheets are considered to require greater care in evaluation. Ice supporting capacity may be increased and ice bridges constructed when necessary, by available techniques. Snow cover is more likely to be present than not and will tend to reduce the rate of ice growth because of its insulating effect. The weight of snow may depress the ice cover sufficiently to cause water to rise through cracks into the bottom of the snow cover.

Tables and charts giving approximate ice thickness requirements for various types and sizes of loadings are available in publications such as the U.S. Army manual *Arctic Construction* (1962). For specific field cases, detailed, individual analyses using combinations of theoretical and empirical knowledge can provide better estimates. In the U.S.S.R. the state of knowledge was greatly advanced during World War II out of dire necessity. This information was summarized in a group of papers edited by B. L. Lagutin (1946). These papers discuss theoretical and simplified methods for calculating ice thickness requirements, the effects of important variables, and factors of safety applicable for various conditions. The factors of safety recommended by these U.S.S.R. investigators vary with the degree of risk that may be justified and the condition of the ice; they range from 1.0 to 1.9, as applied to their equations. A factor of safety of 1.6 is recommended for "normal" ice crossings over ice with only "dry, hair-like cracks."

U.S.S.R. investigators, and many Western Hemisphere investigators as well, have found it convenient for many practical purposes to express the relationship between allowable load and ice thickness by a simplified equation of the following type:

$$P = Ah^2$$

or, incorporating a factor of safety,

$$P = \frac{Ah^2}{F}$$

where P is the total load; h is the ice thickness; A is an empirical coefficient; and F is the factor of safety.

The coefficient A may incorporate the combined effects of various factors, such as type of loading (wheels versus tracks or skis) and temperature conditions. Separate values of A applicable for various special conditions, such as loading at an open crack (free edge), creep deformation of the ice sheet under a stationary load, or travel of a vehicle at the critical velocity (which can cause a stress increase in the ice sheet up to 50 percent), may also be used. Separate coefficients to take specific account of such factors as salinity and temperature conditions may sometimes be added to the equation.

Donald Nevel, ice engineering expert at the U.S. Army Cold Regions Research and Engineering Laboratory,* has suggested that the equations in Table 14.1 may be used to estimate approximate allowable loads for the various conditions indicated.

Gold (1971) found, from a review of a large number of reports of operations on fresh water ice with wheeled and tracked vehicles, that the approximate limit of the reported ice failures could be expressed by the first of the equations in Table 14.1, although good quality ice cover was also used successfully for loadings up to $P = 250\,h^2$. Gold's equation, representing the failure experience boundary, contains no specific

TABLE 14.1.
Ice load-carrying capacity of formulas [a]

Condition	Formula	Source
Wheeled and tracked vehicles on freshwater ice	$P = 50\ h^2$	Gold (1971)
Vehicular loading at a free edge	$P = 25\ h^2$	Nevel [b]
Stationary (parked) load	$P = 25\ h^2$	Nevel [b]
Vehicle traveling at critical speed	$P = 33\ h^2$	Nevel [b]
Vehicle traveling under warm (thawing) conditions	$P = 25\ h^2$	Nevel [b]
Vehicle traveling over old sea ice (1 year or more old)	$P = 25\ h^2$	Nevel [b]

[a] Without factor of safety; values are in pounds and inches.
[b] Personal communication.

*Dr. Nevel's assistance in providing review and suggestions concerning these paragraphs on ice load-carrying capacity is gratefully acknowledged.

factor of safety, and the same is true for the other equations in the table. Therefore, reasonable factors of safety should be applied in using these equations. It is of interest that if a factor of safety of 1.2 is applied in Gold's equation, load values obtained correspond closely with those given by theoretical equations published by the Frost Effects Laboratory (1947) for lake ice with a flexural strength of 150 psi (1035 kN m^{-2}) and a factor of safety of 2.0. This Frost Effects Laboratory theoretical approach was the first adaptation to floating ice of H. M. Westergaard's elastic theory analysis for stress at the interior of a rigid pavement slab. In this analysis, a factor of safety of 1 corresponds with the occurrence of the first crack in the ice, rather than the failure condition boundary of Gold's equation. Under ordinary relatively short-term loading, a considerably higher load is required to cause failure in an ice sheet than is required to cause the first crack. Charts for the bearing capacity of ice under wheeled aircraft, prepared by the Frost Effects Laboratory by its approach, identified in various publications as 'Elastic Theory Analysis,' and employing a factor of safety of 2.0, were used successfully in military aircraft operations using ice surfaces in the arctic beginning in 1947.

Because operation of aircraft from ice surfaces usually involves better defined and less intensely repetitive loadings than may be experienced with surface vehicles, such as trucks and tracked equipment, a minimum factor of safety of 1.2 may be considered appropriate when the equations in Table 14.1 are used in connection with aircraft landings. For relatively intensive use of wheeled and tracked surface vehicles, however, it would appear appropriate to use a factor of safety of 1.6, applicable under the U.S.S.R. criteria for "normal" operations of wheeled and tracked vehicles on excellent quality ice. The more conservative factor may be especially applicable if there is any uncertainty in the basic data, such as concerning possible variations from the assumed ice thickness.

Where practical, snow cover should be removed or compacted. It should not be pushed into windrows at the sides of the traffic lane. Care should be taken to observe any progressive changes in the character of the supporting ice sheet, and special care should be observed if the air temperature rises above 30°F (approx. −1°C). Suspension of operations may be necessary if the air temperature rises well above freezing or the ice begins to deteriorate noticeably.

The above information on the use of ice surfaces is highly condensed and simplified. If the use of ice surfaces for significant surface transportation or for plane landings is seriously considered, detailed guidance on safe procedures can be obtained from those organizations in the United States and Canada that specialize in cold regions technology.

Sometimes supplies, equipment, and materials may be stockpiled at convenient intermediate locations during the summer, then transported the rest of the way during the winter. In remote locations, the most practical means of direct access in summer may be by float plane or helicopter or by water transportation on lakes, rivers, or sea; float planes and water craft are limited to the period when the water is ice-free. Sometimes a small, wheeled plane may be landed on a gravel bar or beach. In all these cases, the size and weight of equipment that can be transported, together with personnel and supplies, may be severely restricted. Light aircraft equipped with skis can land on snow-covered lakes and rivers during a substantial part of the winter, and heavier planes, including ski-equipped C130 aircraft, can use ice landing surfaces during the latter part of the winter. Wheeled aircraft may require snow removal if the snow cover is substantial. Ice will require several weeks after the start of the freezing season to become thick enough to support substantial loads and may remain usable for about 3 weeks after the start of the thawing season before becoming unsafe. Ice thicknesses may reach their maximum values of 4 to 8 ft (1.2 to 2.45 m) approximately between April 1 and June 1. Detailed information on ice thickness observations at numerous locations in North America is available in publications by Ryder (1953, 1954), Bilello, and Bilello and Bates (see References).

During the period in spring when the ice cover is softening and breaking up and during the period when the ice cover is forming and developing initial thickness in the fall, neither boats nor water or ice-supported planes can be used for access.

If especially heavy equipment is needed at the site of the field investigations, such as for installing and loading test foundations piles, it may be necessary to arrange for shipment to the site a year or more in advance. If there is a contractor who will be doing other work in the vicinity during the same time, it will often be advantageous to arrange with this contractor for accomplishment of work requiring heavy equipment or special construction-type expertise.

A generalized planning timetable for a typical engineering soil investigation for a large permanent installation in an undeveloped area is shown in Table 14.2.

Working conditions

Schedules should take into account the seasonally varying working conditions at the location of the investigation. These should include such

TABLE 14.2.
Typical timetable

Operation	Season
Site selection	
Preliminary and reconnaissance studies	Spring
Localization of tentative sites	Early summer
Selection of best site or route	Midsummer
Ground survey and subsurface exploration	
Preliminary survey	Late summer
Detailed subsurface investigations	Early winter

weather factors as air temperature, wind, precipitation, snow cover, visibility, cloudiness, fog, and blowing snow, as well as available daylight and twilight.

Weather records and climatic summaries for specific weather station locations in the arctic and subarctic may be obtained from the U.S. Weather Service, NOAA, Washington, D.C., United States, the Atmospheric Environment Service, Environment Canada, Downsview, Ontario, Canada, or the corresponding meteorological agencies of other countries. Summary information is given in a number of publications, such as Thomas's *Climatological Atlas* (1953). Excellent reviews of available information on climatic conditions in the Northern Hemisphere have been presented by Wilson (1967, 1969). Her monographs not only summarize selected data but also present an excellent selected bibliography at the end of each major section, which lists sources of more detailed data.

For general temperature guidance, Fig. 14.1 shows monthly temperature distributions at selected Northern Hemisphere stations, including absolute maximum, mean maximum, mean, mean minimum, and absolute minimum values. Figures 14.2 and 14.3 show the mean dates of beginning of the thawing season and freezing season, respectively, that is, the seasons when the average daily temperature is generally above or below freezing. If work is planned for the summer months, the difference between these dates at a given location will indicate the approximate length of the thawing season. Actually, where the insulating effects of snow or vegetation are small or negligible and significant heating from radiation occurs at the ground surface, thawing of the ground may start in the spring when the average daily air temperature is still several degrees below freezing. This effect may be observed in Fig. 4.11. This may

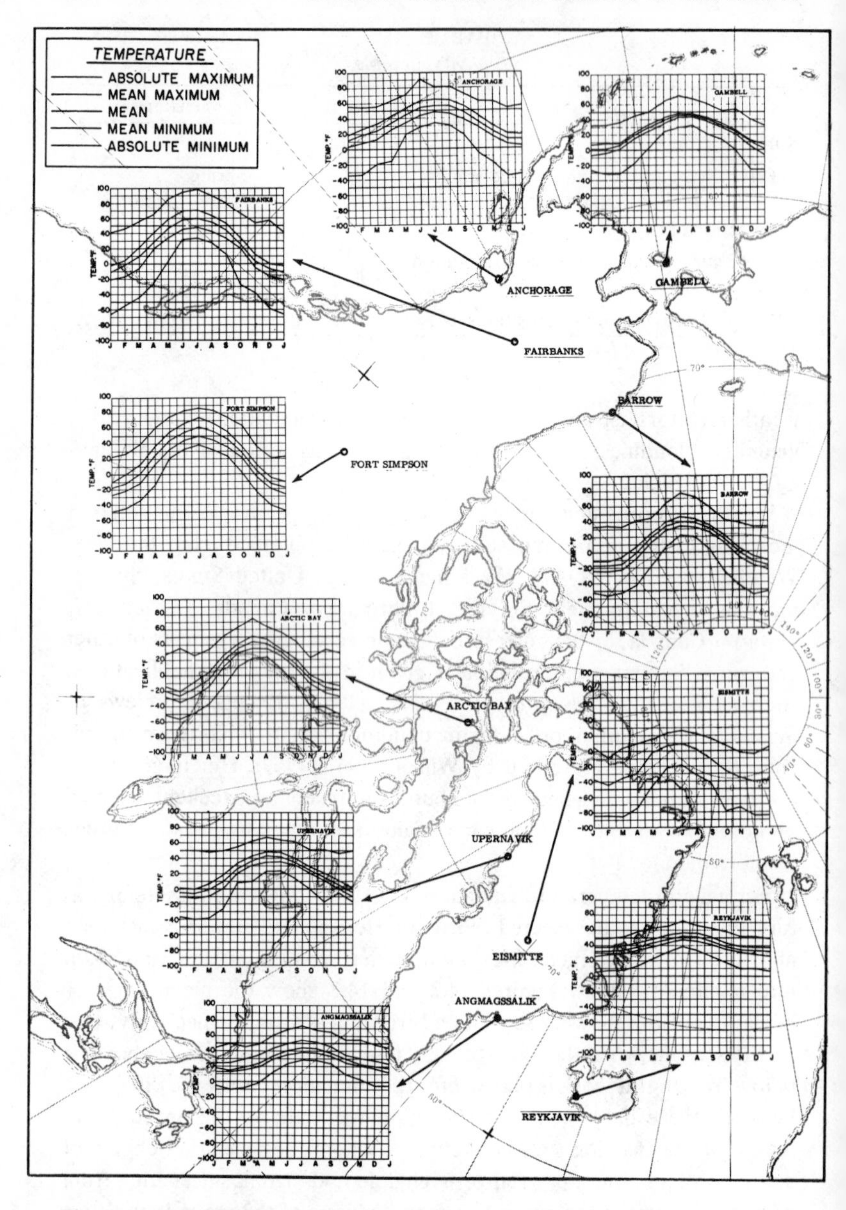

Fig. 14.1. Monthly temperature distributions at selected stations: (a) absolute maximum; (b) mean maximum; (c) mean; (d) mean minimum; (e) absolute minimum (U.S. Navy Weather Research Facility 1962).

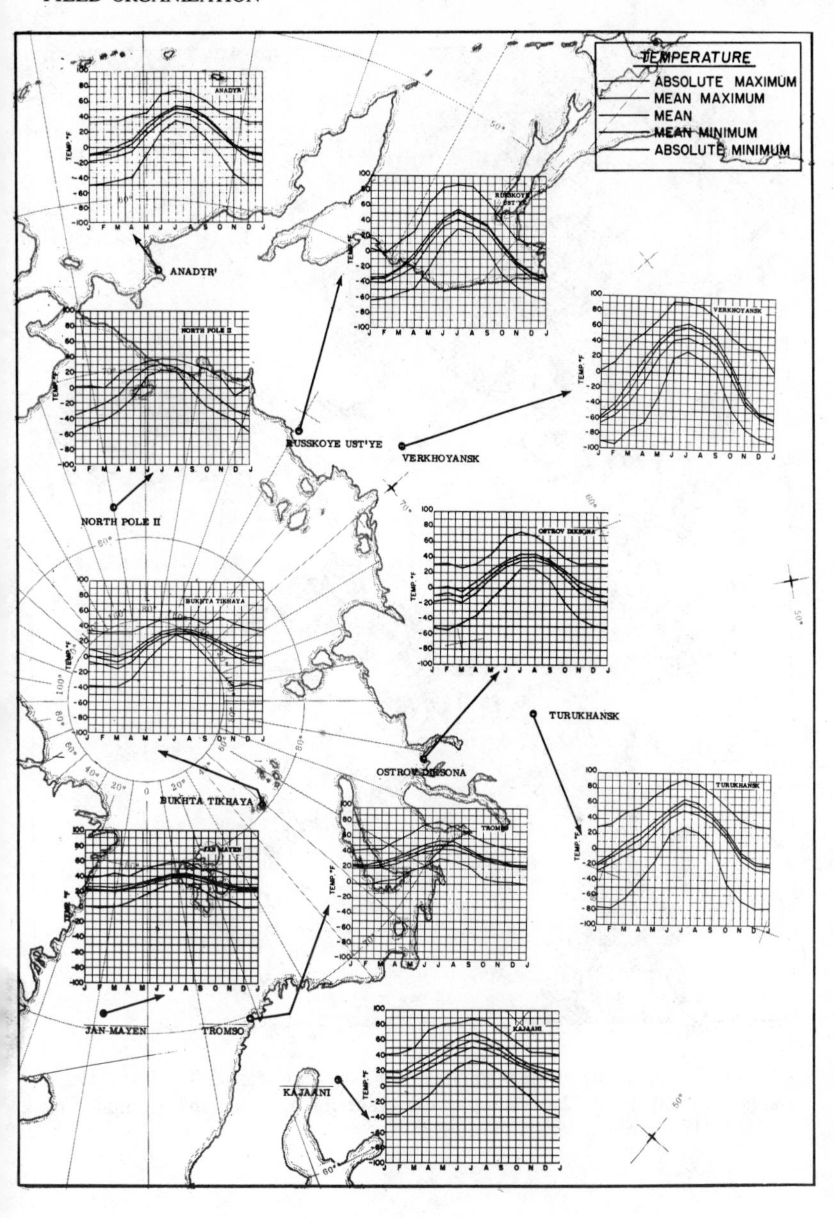
TEMPERATURE
ABSOLUTE MAXIMUM
MEAN MAXIMUM
MEAN
MEAN MINIMUM
ABSOLUTE MINIMUM
ANADYR'
RUSSKOYE UST'YE
VERKHOYANSK
NORTH POLE II
BUKHTA TIKHAYA
OSTROV DIKSONA
TURUKHANSK
JAN MAYEN
TROMSO
KAJAANI
TEMP. °F
J F M A M J J A S O N D J

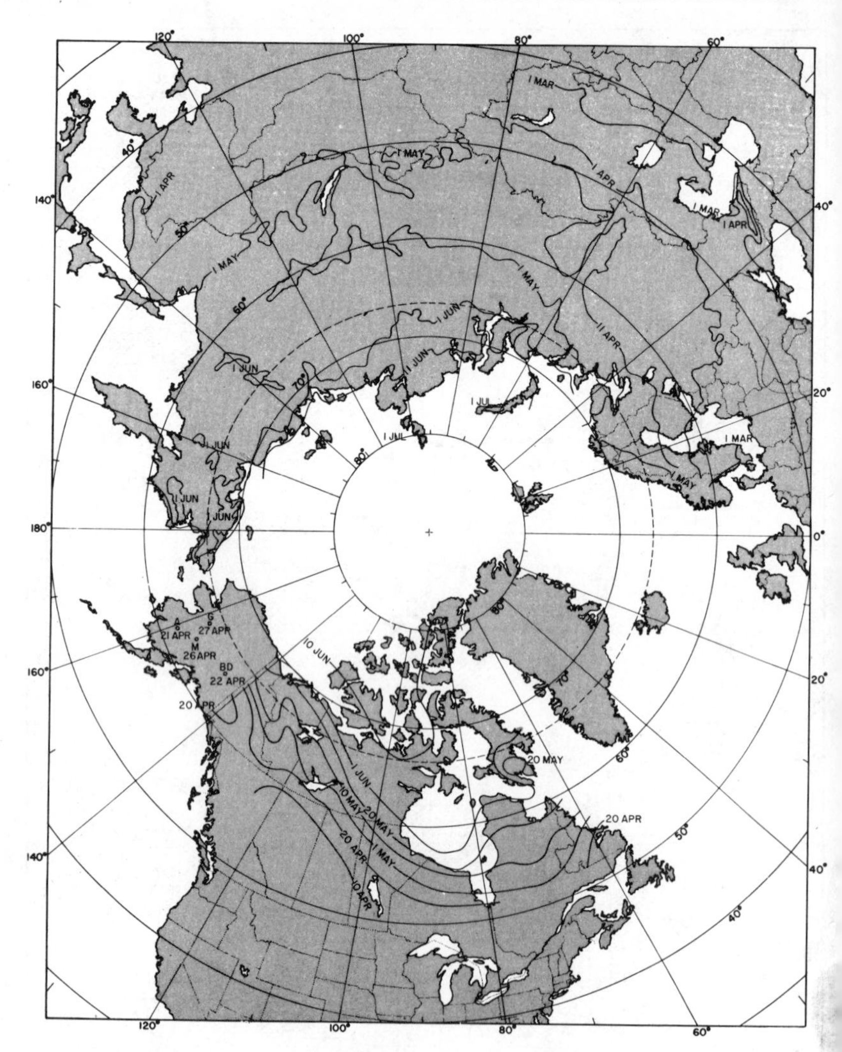

Fig. 14.2. Mean date of the beginning of the thawing season. The latest date is approximately 10 to 12 days after the average for most stations. Compiled by Wilson (1969) from a variety of sources.

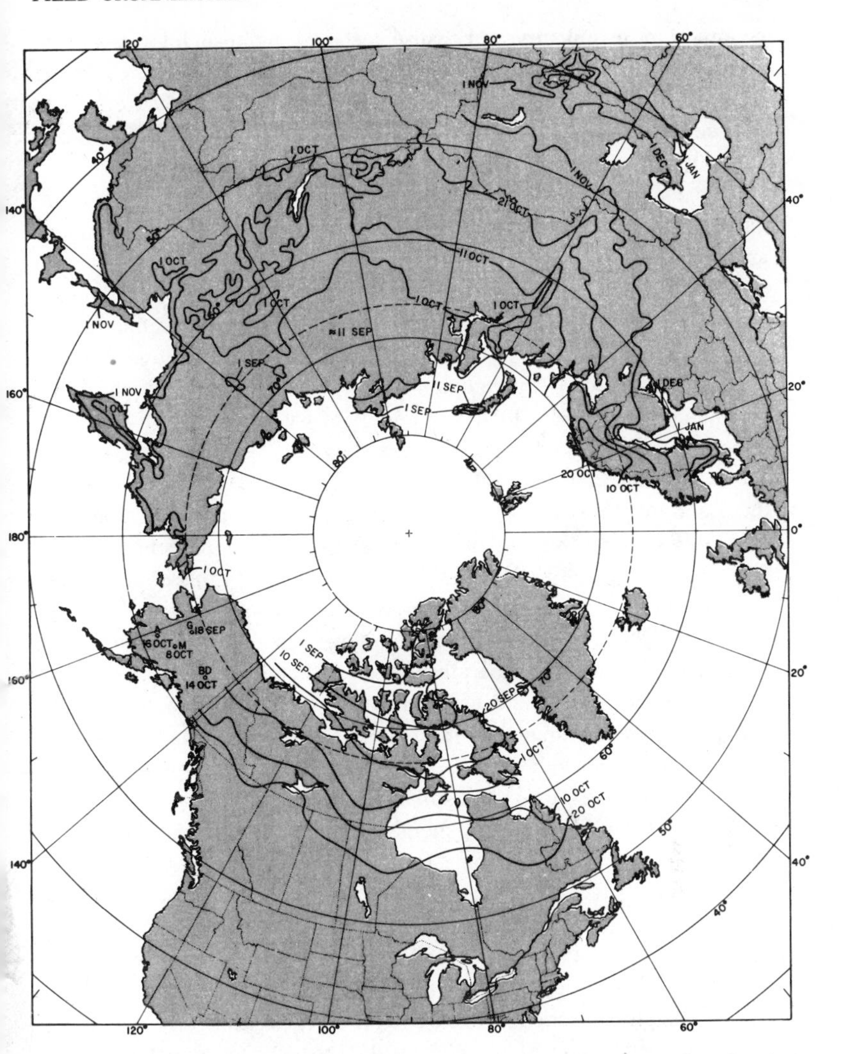

Fig. 14.3. Mean date of the beginning of the freezing season. The earliest date is approximately 10 days before the average for most stations. Compiled by Wilson (1969) from a variety of sources.

be important in soil surveys because surface soils in such locations can be examined in an unfrozen state earlier than would be predicted on the basis of air temperatures. It is also important because of its effect on surface trafficability and because removal of thawed soil by scraping in construction operations can begin a little earlier in the summer.

In many parts of the arctic and subarctic, wind speeds are usually quite low. In and near mountain areas strong winds are more common, and wind speeds up to hurricane velocities may occasionally occur in such locations. The combined effects of temperature and wind are measured by the *wind-chill factor,* which is the measure of the quantity of heat the atmosphere is capable of absorbing within an hour from an exposed surface of 1 square meter (10.76 sq ft). Wind chill values may be estimated using Fig. 14.4. The effects of various levels of wind chill upon human comfort and activities are shown in Table 14.3. Figure 14.5 shows average January wind chill values for North America; note that these are not maximum or extreme values. Low temperature alone can decrease the effectiveness of personnel by about 2 percent for each degree below 0°F (−17.7°C). Any wind will intensify this effect. These considerations must be taken into account in choosing field clothing, providing field quarters, and planning work schedules. To counter the adverse effects of low temperatures and wind chill on people, it is practice in some activities, such as mining, to use mechanization of outdoor activities to the maximum possible extent, thereby reducing the numbers of people who may be affected, and to provide heated enclosures so far as possible for those who must still be outdoors.

Annual precipitation is in general very light throughout the arctic, greater amounts falling in some parts of the subarctic. About half the precipitation may fall as snow, and occurrence and amount of snow cover may be a consideration in scheduling field work. Figures 14.6 and 14.7 show the mean annual first and last dates of seasonal snow cover in Canada. The mean depth of snow at the end of the month with maximum seasonal depth in arctic and subarctic North America is generally in the range between 10 and 50 in. (0.25 and 1.27 m), except in certain coastal or mountain areas; absolute maximum values may generally run about 50 percent higher. Over large geographical areas, the mean depths at the end of the month with greatest snow depth may not exceed about 25 in. (0.64 m). Wilson (1969) has prepared useful brief summaries of available snow cover information. A report published by the Arctic Construction and Frost Effects Laboratory (1954) presents snow depth information in tabular, graphical, and map form for more than 500 indi-

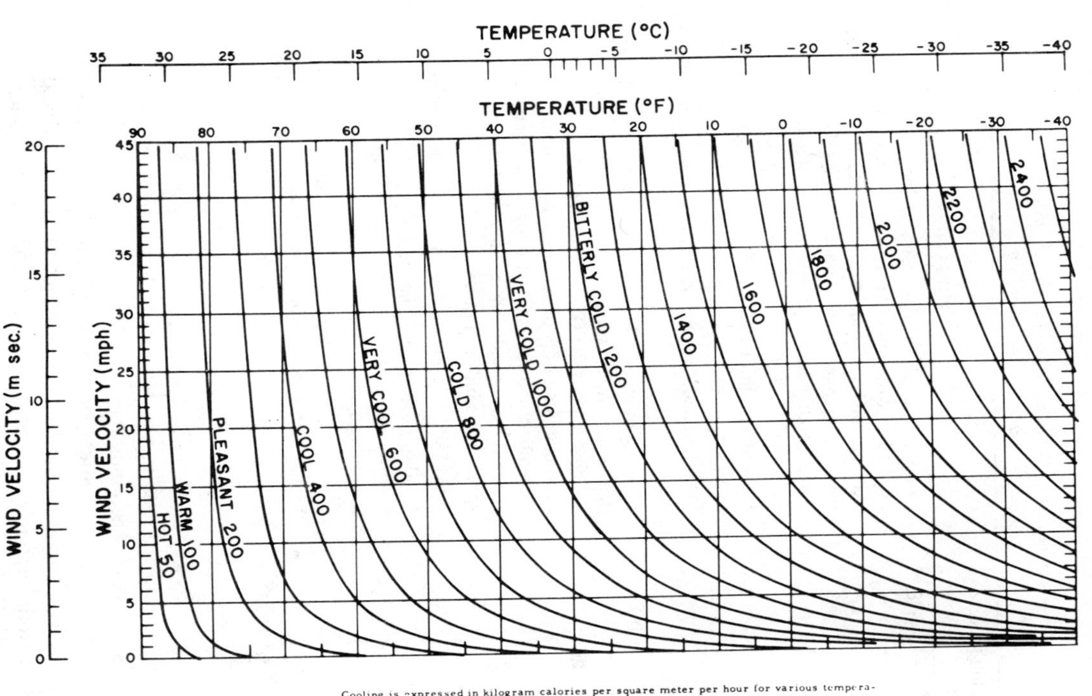

Cooling is expressed in kilogram calories per square meter per hour for various temperatures and wind velocities. The cooling rate is based upon a body at a neutral skin temperature of 33 C. (91.4 F.). When dry cooling rate is less than the rate of body heat production, excess heat is removed by vaporization. Under conditions of bright sunshine, cooling is reduced by about 200 calories. Expressions of relative comfort are based upon an individual in a state of inactivity.

Fig. 14.4. Nomogram of dry-shade atmosphere cooling (U.S. Army/U.S. Air Force 1966a).

TABLE 14.3.
Stages of relative human comfort and the environmental effects of atmospheric cooling

Wind-chill factor kg cal/m²h	W/m²	Relative comfort
600	698	Conditions considered as comfortable when men are dressed in wool underwear, socks, mitts, ski boots, ski headband, and thin cotton windbreaker suits, and while skiing over snow at about 3 mph (1.34 m/s), corresponding to a metabolic output of about 200 kg cal/m^2h (233 W/m^2)
1000	1163	Pleasant conditions for travel cease on foggy and overcast days
1200	1396	Pleasant conditions for travel cease on clear sunlit days
1400	1628	Freezing of human flesh begins, depending upon the degree of activity, the amount of solar radiation, and the character of the skin and circulation. Average maximum limit of cooling during November, December, and January. At temperatures above 5°F (−15°C) these conditions are accompanied by winds approaching blizzard force
1600	1861	Travel and life in temporary shelter very disagreeable
1900	2210	Conditions reached in the darkness of midwinter. Exposed areas of face freeze in less than a minute for the average individual. Travel dangerous
2300	2675	Exposed areas of the face freeze within less than ½ minute for the average individual

Source: U.S. Army 1962.

vidual observation stations in the Northern Hemisphere, using 20 or more years of record where possible.

Blowing snow may also affect surface operations. Mellor (1965) has summarized available information on this problem. Wilson (1969) has stated that from September to June this is a major phenomenon in the arctic and is the chief restriction to visibility.

It is necessary to be alert for local variations in precipitation, snow cover, fog, and visibility, which are determinable only by direct field observations. For example, fog forming locally over cold water bodies in summer may periodically move ashore over landing strips established on the adjacent land areas, even in otherwise clear weather.

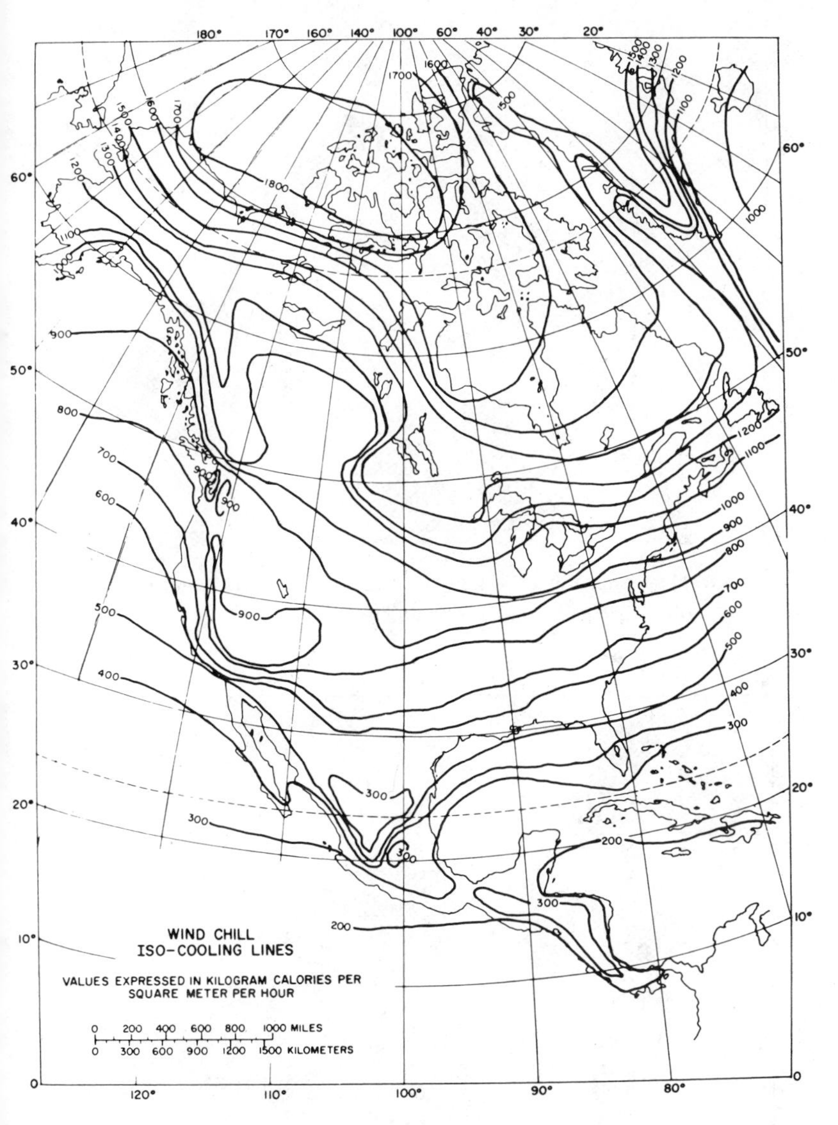

Fig. 14.5. Average January wind-chill values for North America. (U.S. Army/U.S. Air Force 1966.)

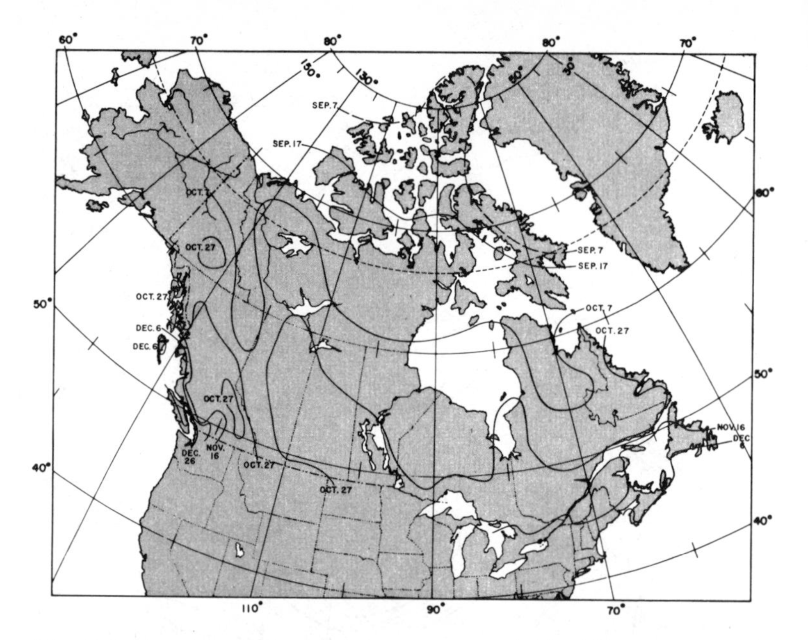

Fig. 14.6. Mean annual date of first snow cover of 1 in. (25 mm) or more in Canada (Wilson 1969 after Boughner and Potter 1953).

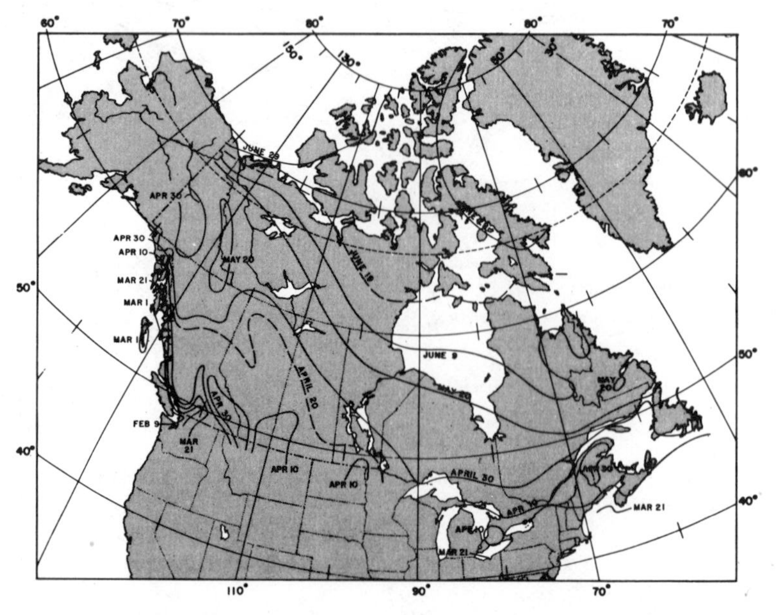

Fig. 14.7. Mean annual date of last snow cover of 1 in. (25 mm) or more in Canada (Wilson 1969 after Boughner and Potter 1953).

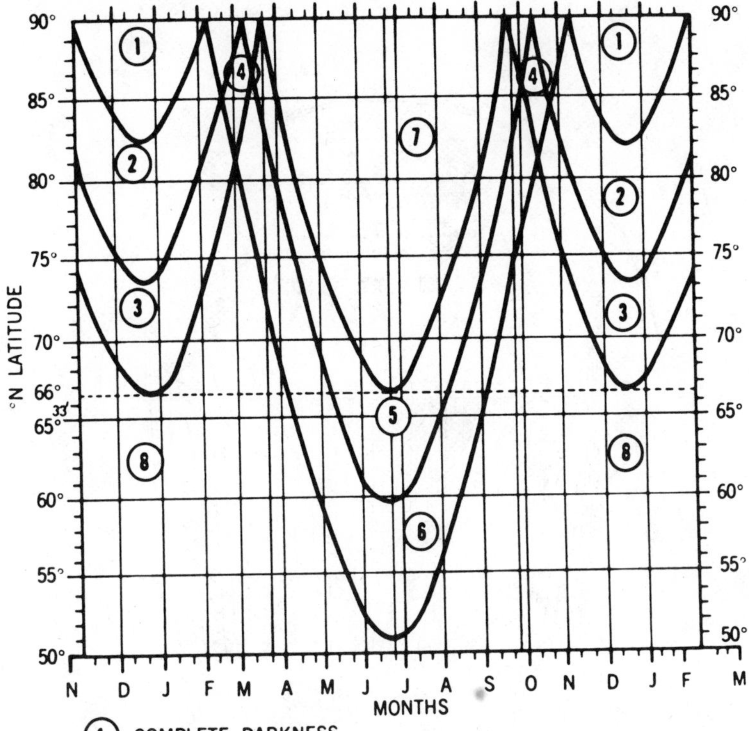

(1) COMPLETE DARKNESS

(2) DARK NIGHT WITH ASTRONOMICAL TWILIGHT AT MIDDAY.

(3) DARK NIGHT WITH CIVIL TWILIGHT AT MIDDAY.

(4) TWILIGHT DAY AND NIGHT

(5) LIGHT NIGHT RESULTING FROM CIVIL TWILIGHT, WITH ALTERNATING DAY AND NIGHT

(6) LIGHT NIGHT RESULTING FROM ASTRONOMICAL TWILIGHT, WITH ALTERNATING DAY AND NIGHT

(7) CONTINUOUS DAYLIGHT

(8) NORMAL ALTERNATING DAY AND NIGHT.

ON THIS CHART IT IS ASSUMED THAT CIVIL TWILIGHT LASTS UNTIL SUN SINKS 7° BELOW HORIZON AND THAT ASTRONOMICAL TWILIGHT LASTS UNTIL SUN SINKS 16° BELOW HORIZON.

Fig. 14.8. Solar illumination in the arctic (Modified from U.S. Army 1962; after Meinardus 1930).

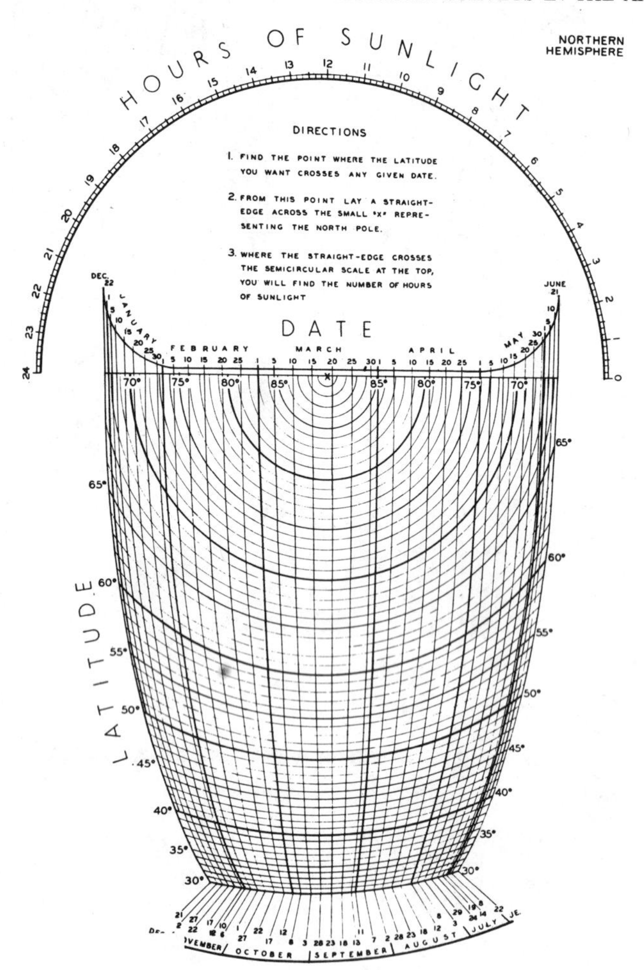

Fig. 14.9. Hours of sunlight (Weyer 1943).

The long hours of daylight and twilight in spring and summer in the arctic and subarctic provide ideal illumination for air and surface operations and long daily hours of outdoor work. Conversely, the long hours of darkness and dim twilight in the winter are substantial handicaps. The combinations of daylight, twilight, and complete darkness which occur between latitude 50° and the pole are illustrated in Fig. 14.8. Note that at 90° latitude there are twice as many days in the year with continuous daylight as with complete darkness. The effect of the large number of

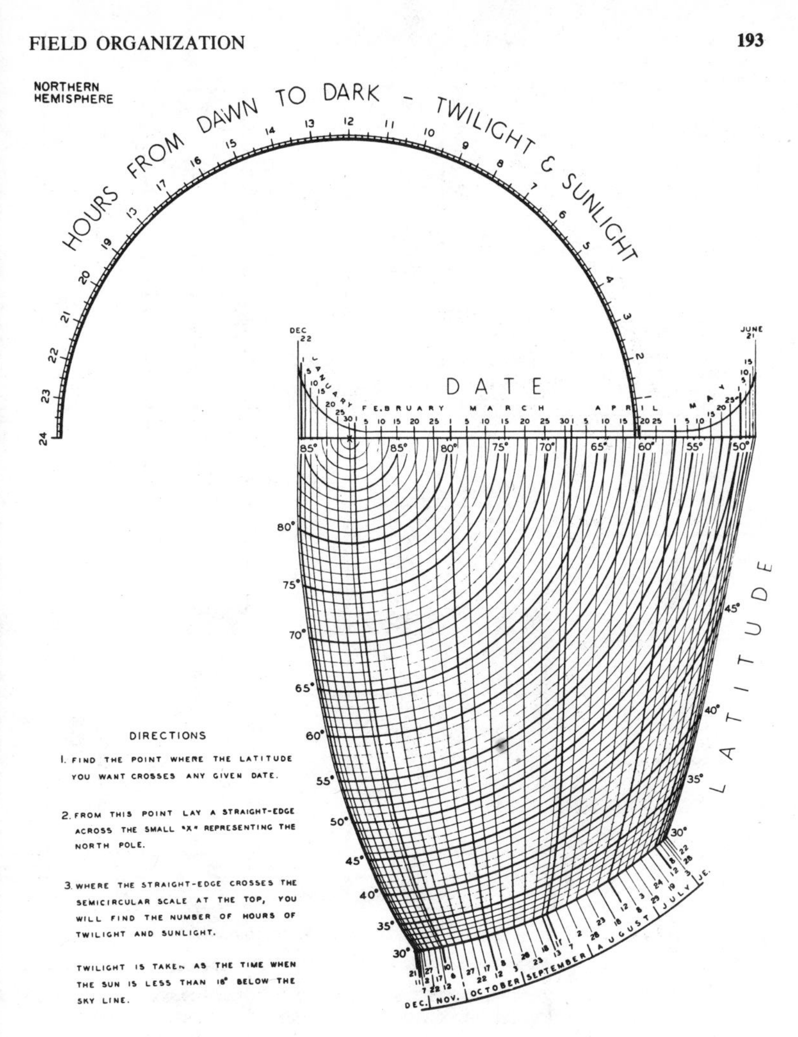

Fig. 14.10. Hours from dawn to dark—twilight and sunlight (Weyer 1943).

hours of daylight is diminished considerably, however, by the relatively low angle of the sun, even at midsummer. The number of hours of daylight or daylight plus twilight may be determined for any day of the year for any arctic and subarctic latitude from Fig. 14.9 or 14.10, as applicable, or from tables of the American Ephemeris and Nautical Almanac (annual). For some winter activities, moonlight may be of assistance during part of each month.

Figure 14.11 illustrates, in the upper part of the chart, the variations of

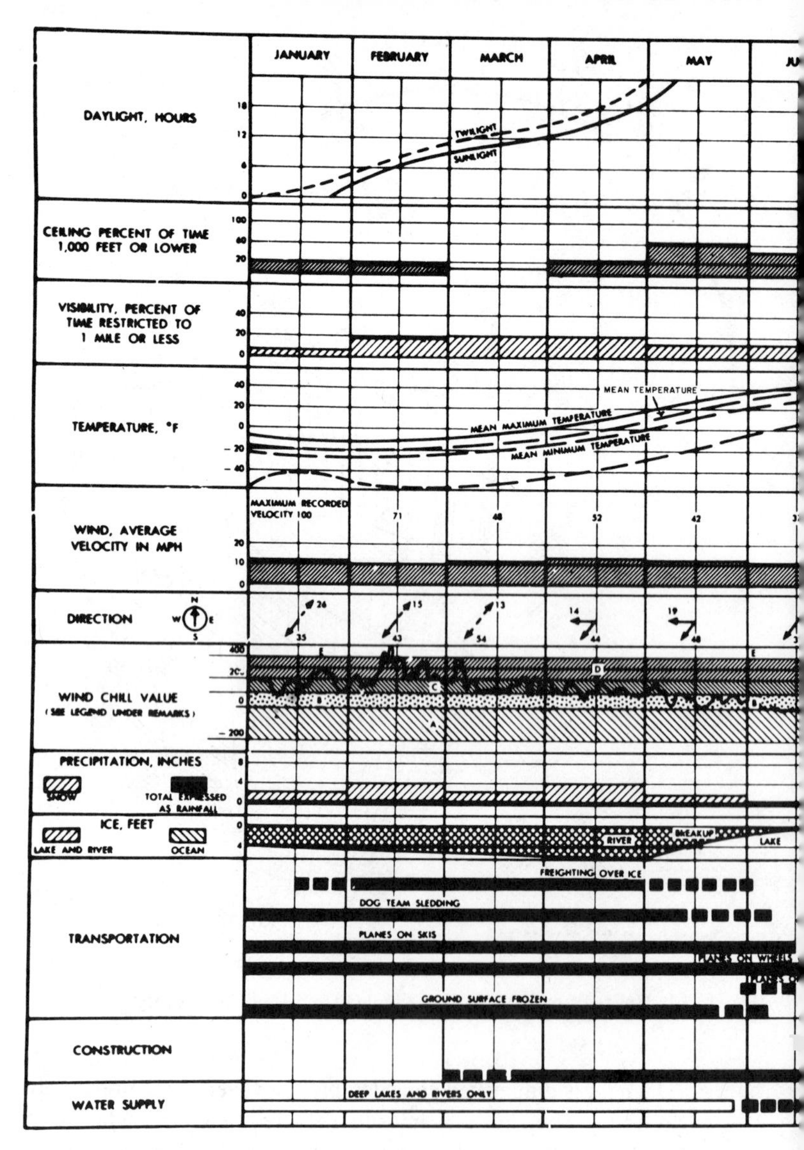

Fig. 14.11. Work feasibility chart, Point Barrow, Alaska (Naval Facilities Engineering Command 1975).

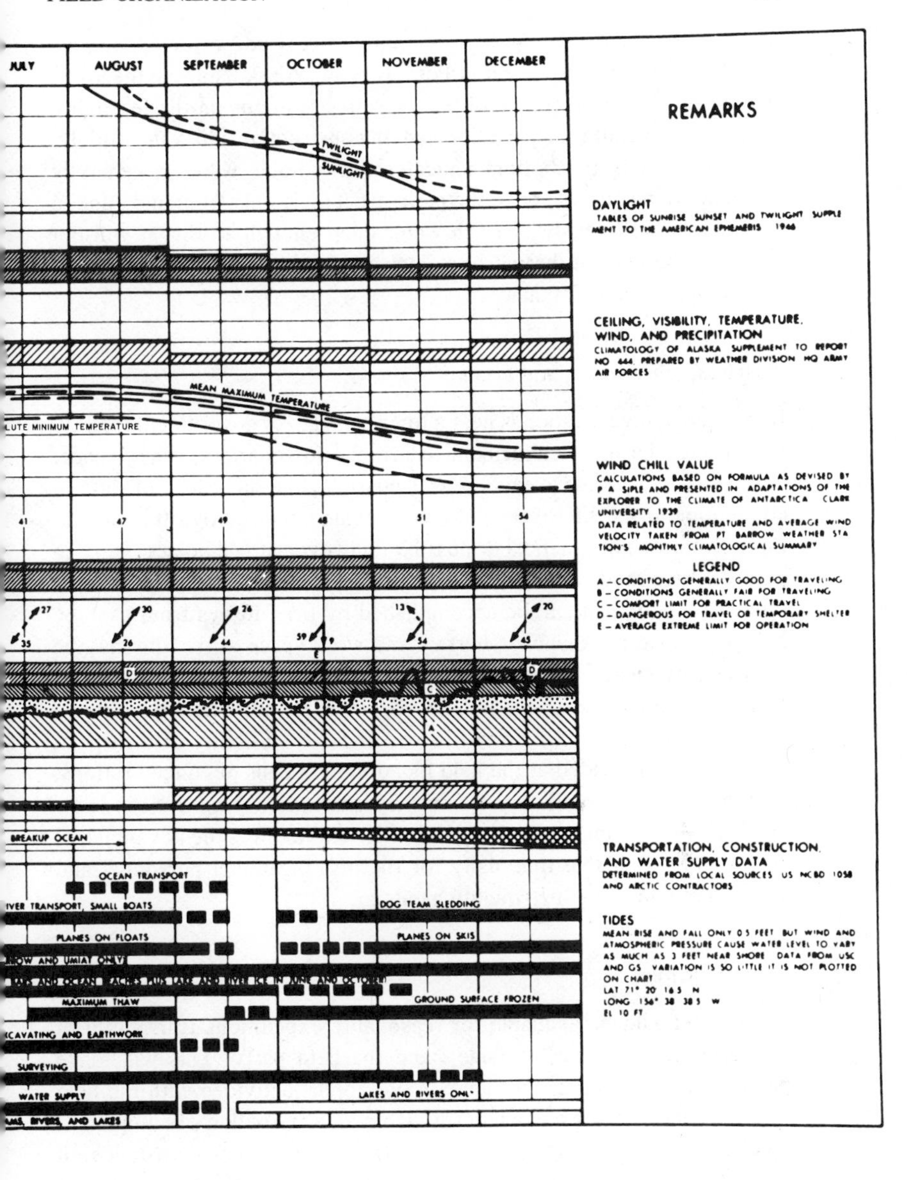
JULY
AUGUST
SEPTEMBER
OCTOBER
NOVEMBER
DECEMBER
REMARKS
TWILIGHT
SUNLIGHT
DAYLIGHT
TABLES OF SUNRISE SUNSET AND TWILIGHT SUPPLEMENT TO THE AMERICAN EPHEMERIS 1946
CEILING, VISIBILITY, TEMPERATURE, WIND, AND PRECIPITATION
CLIMATOLOGY OF ALASKA SUPPLEMENT TO REPORT NO 444, PREPARED BY WEATHER DIVISION HQ ARMY AIR FORCES
MEAN MAXIMUM TEMPERATURE
LUTE MINIMUM TEMPERATURE
WIND CHILL VALUE
CALCULATIONS BASED ON FORMULA AS DEVISED BY P A SIPLE AND PRESENTED IN ADAPTATIONS OF THE EXPLORER TO THE CLIMATE OF ANTARCTICA CLARK UNIVERSITY 1939
DATA RELATED TO TEMPERATURE AND AVERAGE WIND VELOCITY TAKEN FROM PT BARROW WEATHER STATION'S MONTHLY CLIMATOLOGICAL SUMMARY
41
47
49
48
51
54
LEGEND
A – CONDITIONS GENERALLY GOOD FOR TRAVELING
B – CONDITIONS GENERALLY FAIR FOR TRAVELING
C – COMFORT LIMIT FOR PRACTICAL TRAVEL
D – DANGEROUS FOR TRAVEL OR TEMPORARY SHELTER
E – AVERAGE EXTREME LIMIT FOR OPERATION
27
35
30
26
26
44
59
9
13
54
20
45
BREAKUP OCEAN
TRANSPORTATION, CONSTRUCTION, AND WATER SUPPLY DATA
DETERMINED FROM LOCAL SOURCES US NCBD 1058 AND ARCTIC CONTRACTORS
OCEAN TRANSPORT
IVER TRANSPORT, SMALL BOATS
DOG TEAM SLEDDING
TIDES
PLANES ON FLOATS
PLANES ON SKIS
MEAN RISE AND FALL ONLY 0.5 FEET BUT WIND AND ATMOSPHERIC PRESSURE CAUSE WATER LEVEL TO VARY AS MUCH AS 3 FEET NEAR SHORE DATA FROM USC AND GS VARIATION IS SO LITTLE IT IS NOT PLOTTED ON CHART
LAT 71° 20' 16.5" N
LONG 156° 38' 38.5" W
EL 10 FT
MAXIMUM THAW
GROUND SURFACE FROZEN
CAVATING AND EARTHWORK
SURVEYING
WATER SUPPLY
LAKES AND RIVERS ONLY
AMS, RIVERS, AND LAKES

such factors as daylight, visibility, temperature, wind, windchill, and ice thickness throughout the year at Point Barrow, Alaska, and, in the lower part of the chart, the effects of the varying conditions on the feasibility of various forms of transportation and on construction and water supply. The black bar at the bottom indicates that potable water is generally available from sources above the permafrost table only from mid-June to 1 September. If a body of fresh water deep enough so that it will not freeze to the bottom during the winter can be found, however, a year-round supply can be obtained.

Quarters, office space, and meals

If the area of investigation is near a town, village, or government facility, it will often be most convenient to arrange for quarters, office space, laboratory space, and meals there if possible, the daily loss of time in commuting to the site of the field work being balanced by other advantages. If this is not practical or possible, however, it will be necessary to provide these fundamentals at the work site, unless the project is of such limited scope that it can be accomplished by brief forays from base by helicopter or other means. Quarters and work space at the site may be provided by means of tents, Jamesway or Atwell huts or their equivalents, prefabricated rigid buildings, or even mobile-home-type structures, depending on the circumstances and the feasibility of transport. Meals can be provided at the field location by means of canned rations, such as C, E, or K military rations, by sharing cooking duties, or by employing a full-time cook. The latter alternative offers the advantage of saving much valuable time daily for the investigational personnel and can ensure uniformly high quality meals.

Communication

A means should be available for reasonably expeditious transmittal of messages between the "outside" and the field party. This will make possible the continuing coordination of support activities with the field activities and provide a channel through which resolution of equipment, medical and other problems, and calls for emergency assistance, if needed, can be handled. At one extreme, no special communication equipment or arrangements may be needed, as for individual, 1-day trips. If the field location is within reasonable driving distance by road to established telephone or radio communication facilities, special communication arrangements may again be unnecessary. At the other extreme, if the work area is in a very remote location, on terrain lacking topographic features easily distinguishable from the air, and accessible only by air, it will be necessary to provide positive systems for communication between the site and a base area, for assisting air navigation to the site, and possibly for communication with remote units of the field party and obtaining radio time checks.

Safety and medical problems

The types of safety and medical problems that may occur during the field activities should be visualized, and measures should be taken both to minimize the risk of such problems arising and to establish specific courses of action to be followed in case emergency situations arise. In some circumstances, special survival equipment caches may be needed for safety or personnel may need to be divided into self-supporting units for survival purposes. Personnel should be trained in protective measures which may be required against such hazards as fire, animal attack, breaking through ice, immersion in water at low temperatures, severe weather conditions, and equipment-related injuries. Fire can be an extreme danger under adverse weather conditions; even if there is no direct injury, destruction of quarters, clothing, food, and communications by fire can leave personnel in a very serious situation. Bears can be a real danger in some regions. Traveling over ice is especially risky when it is just forming and during spring thaw; open cracks or thin places hidden under snow cover may allow heavy equipment to break through even in midwinter. Winds up to hurricane velocity can spring up suddenly in some locations, and wind-driven snow and whiteouts can quickly reduce visibility to zero. Because of the bulky clothing worn under cold conditions, operation of heavy equipment, such as drilling rigs, may be especially hazardous. Frostbite is an obvious risk during very cold weather, and personnel should be aware of methods for avoiding it and of procedures to follow if it should occur. Hypothermia, or excessive cooling of the body, can lead to death even if temperatures are not extremely low. Snow blindness and illness can develop in a very short time unless eyes of personnel are properly protected with sun glasses when operating on unbroken snow surfaces in spring and summer when the sun is at relatively high elevation. Lack of protection against insects may significantly lower work efficiency in summer.

Personnel assigned to the field work should be in good health. If the work area will be remote, with medical personnel, facilities and supplies not readily available, or available only through an easily interrupted transportation system, each person should be required to pass a thorough physical examination. Persons who might experience incapacitating illnesses or who are dependent on special medicines, such as insulin, should not be accepted for field assignment.

Basic medical supplies should be provided. A procedure for emergency evacuation of serious cases of illnesses or injury should be developed to be ready for use if needed. A procedure for evaluating safety of camp water supplies should be available if needed. Applicable regulations for handling sanitary and solid waste must be complied with.

Supplies, equipment, and materials

A checklist identifying every item of supplies, equipment, and materials required for the exercise, except personal items, should be compiled beginning with the start of the planning. As the planning proceeds, the list will grow in length and detail, and modifications may be made as plans are refined. Table 14.4 shows an example, from an actual project, of the way in which such a list may be arranged. As the list becomes complete toward the end of the planning process, it can be rearranged to group together all items which will be contained in each packing case for shipment.

Unless quarters and meals can be obtained at a village, town, government facility, or contractor's camp, nothing will normally be available at the field location, and all supplies, equipment, and materials must be purchased, rented, or borrowed and shipped to the site. This includes hundreds or thousands of items, such as field clothing; sleeping bags; shelter; field provisions; field stoves; photographic equipment; tools; construction equipment; field exploration, sampling, and testing equipment; sample containers and labels; survey equipment; digging and excavating equipment; reference books; notebooks and writing materials; and miscellaneous items and materials. Heavy items, such as construction equipment and materials, if needed, and fuel, will normally be sought from sources as near the field site as possible.

Each individual member of the field crew will usually be allowed a fixed weight of personal baggage, for which he should have his own individual checklist. This will include personal clothing, toilet articles, tickets and travel documents, identification, wallet and money, reference books and notebooks, camera and film, watch, sun glasses, stationery

TABLE 14.4

Example of a method of listing needed equipment for a field project (partial list)

Test or Category	No.	Desig- nation	Description	Specifications	Function	Status (1)
			SURVEY EQUIPMENT			
Equipment for Astronomical and Survey Measurements:	1	each	Transit, complete with carrying case, protection hood, magnifying glass, sun shade, lens brush, adjustment pin, plumb bob and string, and solar observation attachment.	Engineers' transit with standard accessories.	Determination of geographic location and topographic surveys. Layout if necessary.	NED
	1	each	Tripod	Standard.	For above transit.	NED
	1	each	Stadia Rod	12-ft. folding rod.	Determination of distances and elevations.	NED
	1	each	Stadia Slide Rule with Case	Standard 10″.	Reduction of slant distances to horizontal.	NED
	2	each	Tape	50-ft., metallic.	Distance measurements.	*
	2	each	Rule	6-ft., folding, wood, inches and sixteenths one side, feet and hundredths other side.	Measuring strata depths in test pits, penetration depths, etc.	*
	2	each	Rule	6-ft., folding, aluminum, inches and sixteenths.	Measuring strata depths in test pits, penetration depths, etc.	*

TABLE 14.4 (*Continued*)

Test or Category	No.	Desig- nation	Description	Specifications	Function	Status (1)
			SURVEY EQUIPMENT			
	2	each	Abney Hand Level	Standard clinometer type with leather case.	Determination of relative elevations, direction of slope of terrain.	NED
			TEST EQUIPMENT			
Crystal Structure (Snow)	6	each	Snow Grain Measuring Cups	Drawing No. SLP-Y43.	Determine size and shape of individual snow grains.	
	6	each	Spatula	Wooden, 6 inches long.	Separate snow grains.	NED
	6	each	Magnifying Glasses	Pocket type, 4 power.	Examine snow grains.	NED
Crystal Structure (Ice and Hard-Packed Snow)	1	each	Polaroid Viewer	One, 4½-inch diam. polarizer and one, 4½-inch diam. analyzer.	Examine crystal structure of ice.	
	1	gal.	Plastic Solution	Polyvinyl formal resin in ethylene dichloride.	Impregnate snow in-place for study of intergranular structure.	Requisition cancelled.
			See also Photomiorography Equipment.	See Photographic Equipment.	Study and record granular and intergranular structure of snow and crystal structure of ice.	

Density (Soft Snow)	6	each	Fixed Volume Sampler	Volume 100 cc, Drawing No. SLP-Y40.	Obtain fixed volume samples in soft snow for the determination of density.	
	3	each	Fixed Volume Sampler	Volume 250 cc, Drawing No. SLP-Y40.	ditto	
	3	each	Fixed Volume Sampler	Volume 25 cc, Drawing No. SLP-Y40.	ditto	
	2	each	Balance (modified to take larger pan, attach samples for immersion density tests, and mount in Ice Mechanics Test Kit carrying case)	Triple beam, capacity 2610 grams, reading to 1/10 gram, with stainless steel pan as per Drawing No. SLP-Y27.	Weight samples, fixed volume samplers, immersed samples and miscellaneous items.	*
	2	each	Balance (modified to mount in Ice Mechanics Test Kit carrying case)	Overhead beam, capacity 2500 grams, accurate to one part in 25,000.	Same as for triple beam balance but with greater accuracy.	
	2	set	Weights	Brass, metric, 1 gram to 2000 grams with wooden case and tweezers.	Used with overhead beam type balance.	
Density (Ice and Hard-Packed Snow)	2	each	Hydrometer	Reading in specific gravity, 0.700 to 0.800, 11 inches long.	Determine specific gravity of oil (at various temperatures) used in immersion method of density determination.	*
	2	each	Hydrometer	Same as above, spares.	ditto	
	2	each	Hydrometer	Same as above but 0.800 to 0.900.	ditto	

TABLE 14.4 (*Continued*)

Test or Category	No.	Designation	Description	Specifications	Function	Status (1)
TEST EQUIPMENT						
	2	each	Hydrometer	Same as above, spares.	ditto	
	2	each	Hydrometer	Same as above but 0.900 to 1.000.	ditto	
	5	each	Hydrometer	Same as above, spares.	ditto	
	4	each	Lucite Jar	Drawing No. SLP-Y10.	Contain oil for density determination by immersion method. Small jar for determining specific gravity of oil. Large jar for immersion of sample in oil.	
			See Density (Soft Snow) for Balances			
	2	spool	Thread	150 yds., linen, No. 12.	Tie samples for immersion in oil.	*
	2	each	Rule, Steel	Length 12″, to read to 1/64″, 1/100″ and 1/2 mm.	Measure sample dimensions.	*
	2	each	Caliper, Vernier	Length 5″, reading to 1/128 inch and 1/10 mm.	Measure cylindrical core samples for density determination and for test sample data.	*

	2	each	Meter Sticks with Caliper Jaws	Standard, hardwood, meter sticks with attachable 1¼″ caliper jaws.	Measure large sample dimensions.	
	1	each	Platform Scale	Capacity 75 lbs.	Weigh large samples.	NED
	2	each	Spring Balance	Capacity 60 lbs.	ditto	*
			See also Equipment for Cutting and Preparing Samples.			
Temperature	2	each	Thermometer with Steel Case	−100°C to +50°C., division to 1°.	Measure temperatures in ice, snow, water, air, oil, etc.	*
	5	each	Thermometer	Same as above, spares.	ditto	*
	2	each	Thermometer with Steel Case	−20° to +105°C., division to 1°.	ditto	*
	5	each	Thermometer	Same as above, spares.	ditto	
	2	each	Thermometer with Steel Case	−50° to +50°C., division to 1/5°.	ditto	*
	24	each	Thermometer	Same as above, spares.	ditto	

(1) Items were purchased unless otherwise noted in "Status" column. * Indicates item is component of Ice Mechanics Test Kit. "NED" indicates item to be drawn from available equipment in N.E.D. Soils, Fdn. and Frost Effects Laboratory.

Source: Adapted from Frost Effects Laboratory 1950.

and stamps, and needle and thread, altogether totalling possibly about 75 items.

Support

It will usually be necessary to arrange with a governmental or commercial agency for transportation, aerial photography, and airborne geophysical flights; communication, storage, and base facilities; housing; or other general support. This may be expensive, particularly for aircraft support, but it is essential for execution of the project.

15. Field procedure: pedology

Pedologic surveys in the arctic involve some special considerations, not only from the standpoint of investigators working under adverse climatic conditions, but also in formulating a meaningful scheme for classifying the soils and for dealing with unique problems in sample collection. The field season generally begins about mid-June at which time the low, wet positions continue to remain snow-covered and completely frozen (Drew et al. 1958). July and August are the most favorable months of the year for field work because at that time the soils have attained their maximum seasonal thaw and, in comparison with earlier months, are somewhat drier. After August the weather usually becomes uncertain for extensive field investigations and there is danger of freeze-up.

During the course of field studies it will generally be learned that experiences gained in surveys of temperate climate soil are not always completely applicable to arctic conditions. This is particularly true in establishing field legends, making soil descriptions, and sampling. Instructions to field parties, as given in such well-known volumes as those of Soil Survey Staff (1951), Clarke (1957), and Tyurin et al. (1959), are helpful within limits but none consider the specific problems associated with arctic conditions. Arctic soils have special properties from a number of standpoints; low temperatures, preponderance of ice-water in the soil matrix, complex soil morphologies, complex soil patterns, and others.

Irrespective of the classification system used, the investigator, will, at an early time, delineate certain soil units on a base map, usually without establishing a precise meaning of the units. Boundaries separating first-, second-, and third-order relief forms commonly serve also as major soil boundaries, as do certain vegetative changes in landscape appearance, among others. Tyurin et al. (1959) correctly stated: 'As a rule, soils replace one another gradually through the disappearance of certain features and the accumulation of others. Therefore the pedologist, while striving for accurate representation of the distribution of different soils forming the soil cover of a given territory, must of necessity be satisfied by a more or less schematic outline.' If one superimposes bedrock and surficial geology maps over topographic maps, especially in conjunction with botanical maps, some general soil boundaries usually become evident. In comparison with topographic maps the use of aerial photographs results in a more accurate location of soil boundaries. Aerial photographs, especially if available in stereo pairs, give not only more detailed information on relief features than do most topographic maps, but also degrees, or shades, of gray recorded on the photograph, or in the case of colored photographs, change in colors. In most instances the appearance of the photograph is a reflection of the vegetative pattern or relief, which can be used to indicate soil drainage and related information. It is also critical to have a knowledge of the Pleistocene history of the area because such information will aid the investigator in approximating the nature of certain soil material, including soil textures and drainage patterns, and may even aid in deciphering the general nature of ground ice features.

Combining the form of patterned ground with the soil type to establish a compound mapping unit is a relatively new concept. If one chooses to follow such a procedure, the use of air photographs becomes especially important because the photographic mosaic will provide an avenue for approximating the distribution of the soil types as well as recognizing the multiplicity of patterned ground forms.

Among the considerations in soil map preparation is the actual mapping of the soil variety (corresponding to soil series or soil type in the United States or the soil genus or soil species in the U.S.S.R. Not only must the soil unit be identified but its distribution, including boundaries, must also be established. *The Soil Survey Manual* (Soil Survey Staff 1951) states that soil boundaries are usually not actually traversed; instead they are plotted from observations made throughout their course in soil mapping. Boundaries between soil types or other mapping units generally coincide with observable surface features such as the foot of a slope, crest of a ridge, edge of a swamp, or margin of an outcrop (Fig. 15.1). In making a soils map one ideally should walk with the slope of the land rather than on the contour. This procedure enables the surveyor to recognize more readily the physical changes with respect to landscape elements. Realistically, however, such a procedure cannot always be followed in the arctic for a variety of reasons. In trying to recognize where soil boundaries are likely to occur, various criteria may be used in combination, viz. (i) change in topography, (ii) vegetative changes, (iii) changes in surficial deposit composition, and (iv) change in the general landscape appearance, among others. Figure 15.2 gives some examples of soil patterns associated with various photo-mosaics. Whereas some changes in soil patterns are distinct, others may be gradual and diffuse,

Fig. 15.1. A wet meadow on the north coast of Greenland surrounded by shallow, rocky uplands. The two major soil units, (1) Tundra in the low-lying terrain and (2) Lithosol in the uplands, are easily discernible by surface appearance.

resulting in somewhat indefinite boundaries. Transitional conditions may at times be so gradual that the lines separating soils are largely inferred. Gradual transitions of soils are particularly exemplified with certain aeolian deposits on undulating terrain of the arctic. In some Siberian locations loessial mantles are common, a condition which investigators refer to as 'covering loams of uncertain origin,' implying that origin of the surface deposits has not been fully established and many boundaries have to be approximated.

In establishing the classification system there should be provision for the extreme and complex changes in soil morphology induced through frost action. These complexities are manifested particularly in gley and organic soils. The pedon concept, as described by the U.S. Department of Agriculture (Soil Survey Staff 1975), serves some useful purpose in most regions of the globe but in the arctic, because of the extreme and

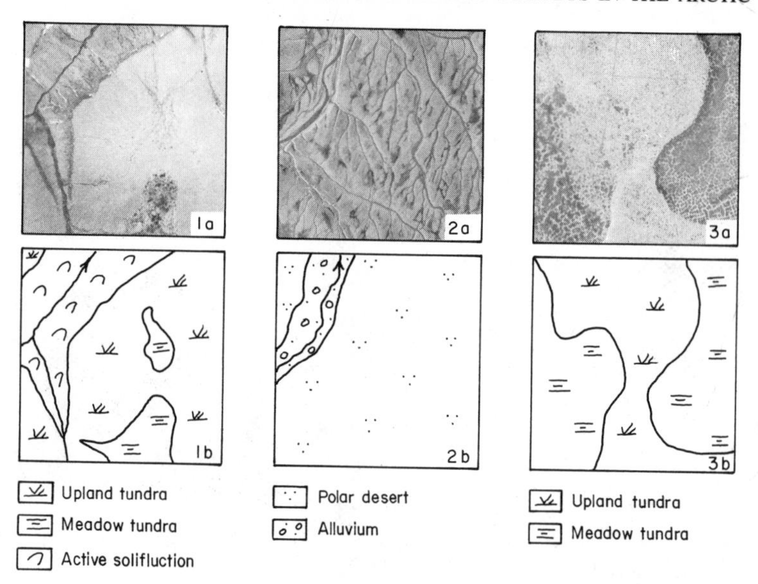

Fig. 15.2. Aerial photographic patterns showing soil distribution in selected arctic areas: 1(a) and (b) aeolian covered outwash of the arctic foothills of Alaska; 2(a) and (b) Beaufort-type sand gravel of Banks Island, N.W.T.; 3(a) and (b) undulating landscape with drained lake basins. Both the original surface and the lake basins (darker areas) are highly polygonized.

unpredictable change in detailed morphology within extremely short distances, this same concept will be severely tested. Fridland (1972), basing his views mainly on the work of such Soviet investigators as B. N. Gorodkov and N. A. Karavaeva, stated that the components of regular cyclic complexes of arctic tundra are genetically interrelated, but at the same time they differ widely in their pattern and composition. Fridland further stated that arctic tundra soil combinations undoubtedly belong to classes of complexes including regular cyclic subclasses, both by the interrelationships of the components and their fairly high degree of contrast.

Plants as soil indicators

The use of plants as indicators of soil conditions has a long history in soil science. When Dokuchaev compiled his global soil map about a century ago, many soil zones were actually projected from vegetative data. Of all climatic regions of the globe, the arctic has some of the greatest possibilities for the use of plants as indicators of soil conditions. The reason is partly because arctic plants have shallow root systems and reflect low-order changes in moisture regimes—especially in the upper part of the soil, a condition which can commonly be detected in the 'gray tone' of the aerial mosaic. There is seldom a high vegetative canopy in the arctic obstructing ground features, but when a 'high' canopy is present it is generally composed of plants such as *Salix, Alnus,* or *Betula,* which indicate rather definite sets of conditions. By arranging plant communities into categories according to moisture preference, micro- and macro-climate, flooding susceptibility, substrate, site quality, and related parameters, it is usually possible to integrate the site factors into a common denominator correlatable with the soil variety. Such correlation has been carried out in Alaska, Canada, Greenland, Siberia, and elsewhere. The use of aerial photographs to identify arctic vegetation and associated landforms including soils has been outlined in the manual *Terrain Evaluation in Arctic and Subarctic Regions* (U.S. Army 1963). This manual indicates the degree of conformity of vegetation to a specific soil condition.

Based on observations in northern Alaska, soil conditions may be correlated with plant associations as follows:

Soil conditions	Plant associations
Xeric, shallow, rocky soils and bedrock with permafrost table about 1 m deep. Permafrost and seasonal frost are generally dry	Barren communities with lichens, mosses, and dwarf heaths. Where bedrock is exposed, crustose lichens and scattered herbs
Deep, mature, well-drained soils. Permafrost table about 1 m deep	Xerophytic mosses, lichens, dwarf heaths, and herbs
Gley (wet) soils (the most common condition). Permafrost table about 30 to 40 cm deep with ground ice	Cottongrass tussocks (*Eriophorum*) and dwarf heaths. Low areas colonized with lowland cottongrass (*Carex aquatilis*)
Organic (bog) soils. Permafrost table about 20 to 30 cm deep, but less where peaty material is well-drained	Lowland cottongrass (*Carex aquatilis*). Where there is shallow, standing water, *Dupontia* is generally present. Where peaty material is relatively dry, lichens and dwarf heaths colonize the substrate

The above information will help one in placing the soils into broad, provisional categories. Depending upon the soil classification system used, the categories can be further divided according to detail.

Floodplains and terraces of the arctic have a variety of soils associated with them. Variations in textural composition, moisture regimes (including flooding), and related site factors dictate a range of plant communities. Dense willow stands commonly colonize the floodplains, but in river wash areas the site may be without vascular plants. In northern Alaska the willow (*Salix alaxensis*) may attain heights up to 8 m. Under such conditions the soil will be well drained to a depth of a meter or more, but will lack significant genetic horizonation. As the vegetation flourishes and the canopy continues to develop, a peaty layer forms, followed by a rise in the frost table and water table. With the rise in the water table there is an accompanying deterioration of the willow community.

Vegetation can also be used as a reliable indicator of soil conditions in the high arctic. Such terms as *desert-glacial, arctic desert, polar desert, rock desert, fell field,* and others have been used for characterizing the dry aspect of the far northern reaches of land. On the dry, windswept uplands the vegetation consists largely of lichens, only a few vascular plants, such as *Poa, Festuca, Luzula, Saxifraga,* and *Potentilla,* managing to attain a foothold in sheltered sites. The wet depressions colonized by *Deschampsia, Dupontia, Arctophila, Carex,* and others generally signify Tundra or Bog soils with very shallow permafrost.

Pedologic classification procedures used by various plenary bodies and individuals (Chapter 6) show that there are areas of agreement in the recognition of certain soil conditions, such as gley soils, shallow soils, organic soils, and so on. There is also considerable agreement as to recognizing cardinal properties of soils, such as organic matter content, pH values, cation exchange data, clay mineral suites, and textural composition, among others, in the characterization of soils. Yet, regarding principles used in classifying soils, opinions tend to be quite diverse. Irrespective of classification procedures used by various nations and individuals, there are usually some equivalent terms within the various systems.

Within the framework of the pedologic classifications (Chapter 6), it is believed necessary to add several new parameters to the final soil unit delineated on the map. In addition to the patterned ground problem, frost action and other geomorphic processes frequently elevate or depress the local land surface so that the 'wetness factor' of the site may change from the original conditions under which it formed. This problem has not received adequate attention. For example, Bog soil is normally water-saturated during most of the period of vital activity. But Bog soil in the arctic may be in an elevated position from cryogenic processes, resulting in the water draining from the local elevations to the depressions. Thus the elevated peaty material is partially drained, and shallow-rooted plants colonize the substrate. Whereas some soils may be elevated into an environment that is drier than that under which they formed, other soils may be depressed into a wetter environment, resulting in more hydric plant communities colonizing the soil. Bog soils in the tundra zone normally have a plant community consisting of *Eriophorum angustifolium* and *Carex aquatilis* as the dominants. If, however, the Bog soils are elevated, dwarf birches and dwarf heaths become established, along with lichens, and mosses. On the other hand, if the Bog soil is depressed so that it is water-covered, the emergent aquatics, such as *Dupontia fischeri,* or floating and bottom-dwelling plants are present (Fig. 15.3). The above principle applies to mineral gley soils as it does to those of an organic character (Tedrow and Cantlon 1958).

Patterned ground

We now turn to the problem of arctic soil classification as it relates to patterned ground. It has been suggested that the type of patterned ground (along with the 'wetness factor') be combined with the soil variety to form a mapping unit (Tedrow 1977). In the arctic, patterned ground is not universally present, but when it is there should be some provision for its inclusion with the soil variety. Drew (1957) prepared a soil map which included the patterned ground form and 'wetness factor' of the site (Fig. 15.4).

Pedologists traditionally make abundant use of road cuts and excavations during field investigations, but in the arctic region such exposures are generally not available. Therefore, one has to consider some form of excavation for soil examination and sample collection. Where river cut banks or landslide sites exist, the soil is usually so disturbed and the true thermal regime so altered that a number of difficulties are introduced. In the shallow, xeric sites of the mountains, as well as sand dunes, very well-drained alluvial deposits, eskers, moraine crests, and other very well-drained sites with a dry permafrost condition, the field procedure may be similar to that of the temperate regions. Under such very well-drained to xeric conditions, permafrost is deep, and seasonal thaw takes place during the early summer months.

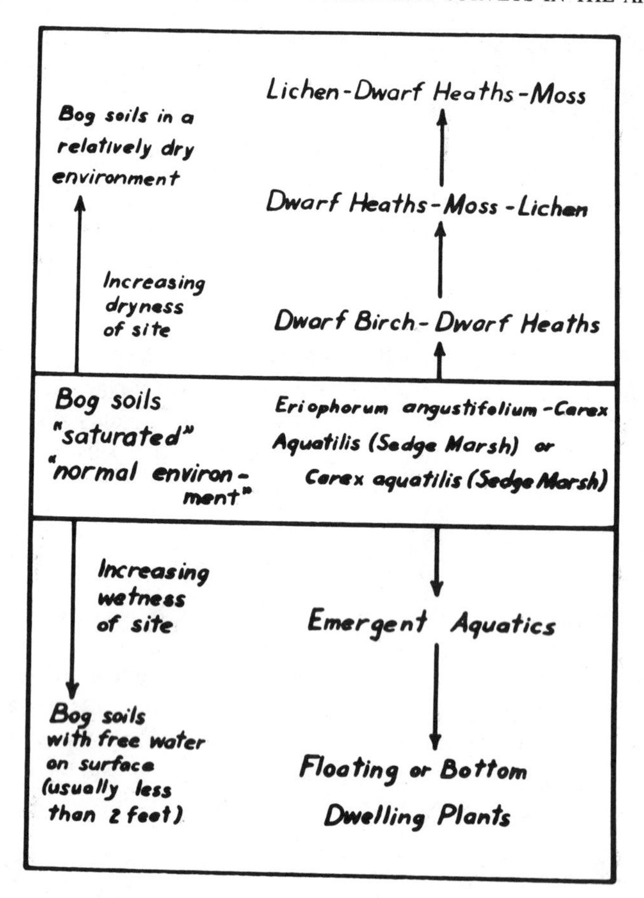

Fig. 15.3. Plant communities associated with Bog soils under various degrees of wetness (Tedrow and Cantlon 1958).

Soil sampling for pedologic purposes

Poorly drained soils of the arctic are among the most difficult to sample anywhere on the globe. Up until some 30 years ago, nearly all sampling was confined to the active layer. This resulted in descriptions and sampling to depths of only 25 to 30 cm, with a few notations about conditions within the underlying frozen strata. Due to this shallow soil sampling, the recorded morphology of arctic soils over the years is strikingly incomplete. Digging soil pits by hand in the frozen sections of Tundra soils

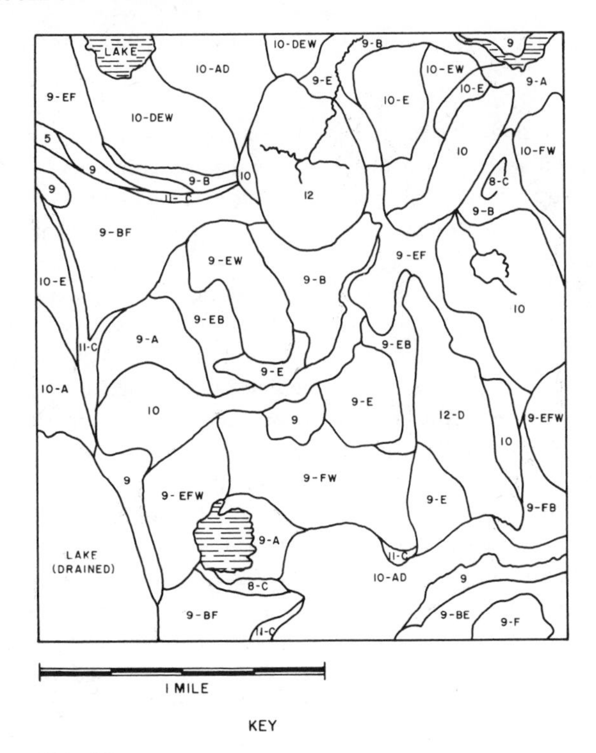

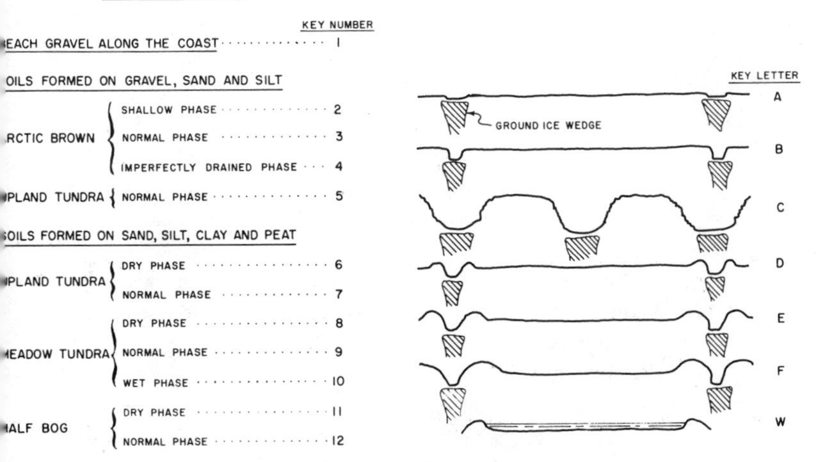

Fig. 15.4. Soil map from Point Barrow, Alaska, area. Shown are the "soil types," degree of wetness of the sites, and dominant polygon forms (Drew 1957).

and Bog soils is most difficult. The investigator will find that once the frozen strata are encountered further attempts at digging are laborious and very slow; the surface waters draining into the sump of the pit compound difficult conditions. Pedologists traditionally make extensive use of soil augers in sampling soils. In the arctic the surveyor will learn that the traditional hand-operated soil auger is useful only in well-drained areas. The conventional soil auger cannot penetrate the frozen soil. Soil sampling tubes with tempered tips (about 1 in [2.5 cm] diameter) have generally proven to be ineffective in penetrating frozen soils.

Mechanical drills have been used with some success. Drilling techniques used in frozen ground vary according to objectives, equipment availability, logistics, and other factors. Frozen soils have been successfully cut with powered disk saws to depths of 34 in (86 cm) (Mellor 1975). Various rotary drill coring techniques have proven to be successful (Hvorslev and Goode 1957; McCoy 1965; Lange 1968, 1973*b*). After frozen cores were removed, Sellmann and Brown (1966) placed them in flexible polyethylene tubing and capped and stored the tubes under refrigeration. Heavier types of hand-held gasoline-driven or pneumatic drills operating on compressed air have proven to be suitable for penetrating frozen soils, but if an air compressor is required the field investigator does not always have total field mobility, and there may be limitations as to complete choice of location of the drilling site. Furthermore, where bouldery till is present, the digging problem becomes rather acute.

Small rotary drills may be used with success if one can carry out the assignment by examination of cores alone. Small-diameter cores are adequate for pollen sampling, radioactive carbon dating, and some general morphological recordings, but the investigator is limited in studying the 'field effect.' Large holes, up to 0.8 m or so in diameter may be augered out using heavy mechanical equipment (Fig. 15.5). This is a satisfactory method of sampling in the frozen layers (Sellmann and Brown 1965).

Explosives have proven to be effective in blasting open pits for soil examination and sample collection. A 2- to 3-cm metal pipe is driven into the soil to a depth of about 60 cm and, after removal of the pipe, a springer charge is placed in the hole and detonated. This is followed by a small charge of conventional explosives, such as TNT, placed in the newly formed hole. After the detonation of the TNT, the blocks of frozen soil can be easily removed from the pit and the investigator will usually have at least an hour or so to work before flooding occurs. There is

Fig. 15.5. Boring a soil pit near Point Barrow, Alaska, using a power rig.

Fig. 15.6. A Tundra soil pit near Umiat, Alaska, dug with explosives.

potential contamination of the soil through the use of explosives. Figure 15.6 is an example of a soil pit excavated with explosives.

Preparation of the soils mapping legend

In preparing a provisionary soils legend for the arctic, it is probably best for the investigator to focus first on the field mapping unit and not to be unduly concerned about the broader classification categories. Most investigators now agree that well-drained soils, plus the shallow, rocky soils and the various gley soils, extend from the tundra zone to the northernmost land extremities and that organic soils also extend well into the high arctic. Even though there are identifying soil patterns ascribable to climatic variations (e.g. Polar desert soil is generally confined to the far northern latitudes), most of the soil taxonomic units are present throughout the arctic. The first generalizations (e.g. shallow soils, bogs, alluvial deposits, dunes, etc.) are usually rather easily identified. Using landforms in conjunction with bedrock geology, surficial geology, and topographic and vegetative maps, one can predict some soil properties; detailed soil characteristics must finally be verified by field inspection. The *modus operandi* for field identification of the well-drained soils and the shallow mineral soils is similar to that of the temperate region. In gley soils and, to an extent, organic soils, however, the frost effects may be so great that some soil features may not exist in a repetitive pattern.

Table 15.1 shows some major considerations during the course of establishing a pedologic soil survey. The exact nature of the information in each column is not specific. Usually the composition of the rock will play an important part in determining the nature of the soil, especially if the rock has some specific identifying characteristics, e.g., a granite giving a coarse-textured residue, as compared with a medium-textured residue from a basalt. If one considers the landform together with the nature of the deposit plus textural composition, especially in conjunction with the vegetative cover, the general moisture regime at the site should be predictable.

With an evaluation of all factors shown in Table 15.1 against a background of soil development, it should be possible to construct a soil legend which will satisfy our taxonomic procedure.

Collection and storage of samples for pedologic purposes

Sampling and storage of soils, especially those from the permafrost, introduces some problems, particularly at sites with a high ground ice

TABLE 15.1.
Some considerations for establishing specific soil series during preparation of arctic soil classification legends

Rock mineral composition	Landforms and surficial deposits	Soil textures [a]
Crystalline rocks (separations according to grain size and mineral composition)	Coarse-grained, stratified deposits (separations based on origin, age and composition)	Coarse-textured soils
		Moderately coarse-textured soils
		Medium-textured soils
Clastic rocks (separations according to grain size and mineral composition)	Nonstratified deposits	Fine-textured soils
	Lake and swamp deposits	
	Aeolian deposits (loess, dunes, etc.)	
Carbonates	Scree deposits	
Gypsum		
Bentonite		

Other considerations:

- Time factor
- Climatic factor (including climatic change)
- Site factor
- Developmental sequences of soils and vegetation
- Drainage—surface and internal
- Frost action
- Relief
- Ground ice
- Plant community
- Thickness of the soil cover

[a] Stony conditions also need to be considered.

content. Brown (1966) determined the 'moisture' content of a soil near Barrow, Alaska, and found that it ranged from 70 to 210 percent. When blocks of frozen material are excavated, they should be placed in plastic bags or other watertight containers. An alternative method is to air-dry samples in the field (a most difficult task) and ship them to a nearby field station where they may be dried at slightly elevated temperatures. In so doing, the soluble constituents of the frozen matrix will accumulate and give abnormally high readings for such measurements as conductivity and possibly exchangeable cations. Brown (1966) showed that the specific conductance of frozen soils at Barrow, Alaska, ranged between 470 and 12 200 micromhos with an increase with depth. Such ions as sodium, potassium, calcium, magnesium, and chlorine increased with the conductivity values. Another potential problem arises when transferring soil from its original frozen state to room temperature. In so doing, some mineralization of the organic materials may occur, minerals may be altered, and the general composition of the soil solution may change owing to the elevated temperatures induced during sample preparation. With well-drained sites the problem of ice content does not pose such a complex problem as it does in the wetter areas with a high content of ground ice. If, however, the well-drained soils are sampled while the soil is frozen, there may also be an excess moisture problem.

To return to the problem of sampling frozen soils for pollen analyses or radioactive carbon content, the samples should be removed from the site with a minimum of disturbance and wrapped in plastic and foil; the sample depth and orientation of the soil should be noted. Pollen samples should be forwarded to the laboratory without delay. In the case of studies involving the preservation of frozen soil fabrics, the natural problems of sampling and transporting the sample in a frozen state must be considered.

There may be times when it is desirable to collect frozen soil and preserve it in the frozen state for future examination. This technique has been used by geologists, glaciologists, palynologists, engineers, and others who need to preserve the ice mineral-organic fabric. Not only does the process involve preserving the frozen soil in a dry ice pack during transport, it also requires refrigerated storage after the soil arrives at the laboratory. The cooling of soil samples with dry ice may alter the specimens, particularly with respect to the nature of soil fabric.

Micromorphological examination of arctic soils has been only fragmental. Theoretically though, the techniques can be used to establish the composite picture of soil formation. While collecting samples for future analyses, the investigator can use conventional methods for the well-drained sites (Kubiëna 1970). Those samples with a high ice content pose special problems, however, because they should be preserved in their original frozen state. Once the ice melts gross deterioration results and obliteration of the original fabric. To overcome this difficulty those samples with a significant ice content need to be preserved in a continuously refrigerated condition (Gow and Williamson 1976).

16. Terrain evaluation

An essential feature of site or route evaluation procedures is the analysis of terrain features and details. If an understanding of the general surficial geologic and hydrologic details, the bedrock geology, and the physical geography of the area is developed, together with knowledge of the arrangement of soil and rock strata in relation to topography and of past and continuing geologic and terrain-modification processes, a rational basis will be provided for interpreting subsurface information for engineering purposes.

Only soil-related aspects of this approach are considered here.

Soil varieties and their landforms

The major soil varieties may be grouped under four general headings: (a) water-deposited materials, (b) glacial deposits, (c) wind-laid deposits, and (d) materials formed from on-site bedrock. Soils in each of these groups or their subgroups tend to be characterized by certain typical natural surface features, have particular properties which affect their behavior in the arctic and subarctic environment, and tend to be affected differently by the environment in terms of relief, drainage patterns, erosional and degradational features, vegetative cover, special surficial markings, and other features. Water-deposited materials may be alluvial, lacustrine, or marine. Four broad categories of glacial deposits are moraines marginal to glaciers, ground moraine, kettle-kame deposits, and eskers. Wind-laid deposits include sand dunes and loess. Materials derived on-site from bedrock include residual soils and colluvial materials (detritus, talus, scree). Deposits such as outwash plains, kamelike terraces, eskers, coarse-textured moraine, and sand dunes, usually easily recognizable by their surface forms, are commonly, though not always, free of large quantities of ground ice. Drumlins and fine-grained moraine and alluvial deposits, on the other hand, tend to contain considerable ice when frozen. Unmodified loess may frequently contain little ice, but when the materials have been reworked by downslope movement they may contain massive ice deposits. Topographic position is an important factor affecting the suitability of soils for construction projects. Soils in elevated, well-drained positions are likelier to provide more satisfactory construction sites than those in low, poorly-drained locations. For the most part, landforms such as sand dunes, eskers, till plains, and kettle-kame deposits may not be greatly altered by the presence of permafrost, except for surface details. Nevertheless, permafrost is an essential feature in the development of such special features as pingoes and thermokarst lakes or depressions.

Groundwater and surface drainage

Those who conduct subsurface investigations in permafrost areas should be aware of the variety of subsurface positions in which groundwater may be encountered. During the thawing season, one tends to find free water in the active layer, often saturating the layer to the ground surface. As indicated in Fig. 8.16, free water in this layer tends to disappear as freeze-back progresses in the fall. If a residual thaw layer is present, free water may be present therein year-round. Unfrozen strata containing saline water are not uncommon within permafrost, especially in coastal areas, though they are also encountered elsewhere. Groundwater is also encountered under permafrost, sometimes under artesian pressure, as illustrated in Fig. 8.17. When artesian flow from below permafrost is encountered in wells, care is required to avoid uncontrolled flow to the surface on the outside of the casing, as this water will have a significant capacity to thaw permafrost (Linell 1973). Subpermafrost water tends to have very high levels of dissolved solids and gases and may, therefore, not be potable. Small and large springs whose sources are below the permafrost may be found. In some locations thermal springs occur.

General drainage patterns are determined primarily by the nature and structural features of the subsurface materials. Dendritic, radial, annular, trellis, and rectangular drainage patterns are recognized, each commonly reflecting a particular kind of underlying bedrock. These are illustrated and described in Fig. 16.1. The characteristics of the soil mantle and the relative hardness and structure of the bedrock determine the details of the drainage patterns. Where surface flow causes underground ice to thaw or erode, distinctive drainage-related surface features develop. For example, 'button drainage' or beaded drainage, consisting of a series of pools connected by surface drainage which may be sluggish, as shown in Fig. 16.2, is produced by local melting of ice and is a definite indicator of permafrost containing masses of ground ice. Arctic and subarctic soils tend to be eroded very easily when exposed to running water.

(a)

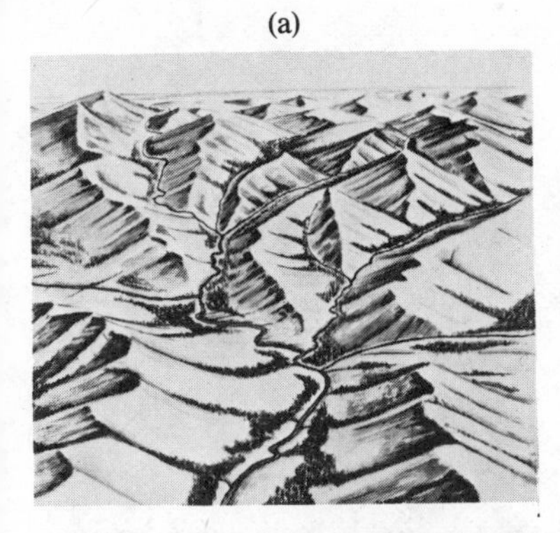

(b)

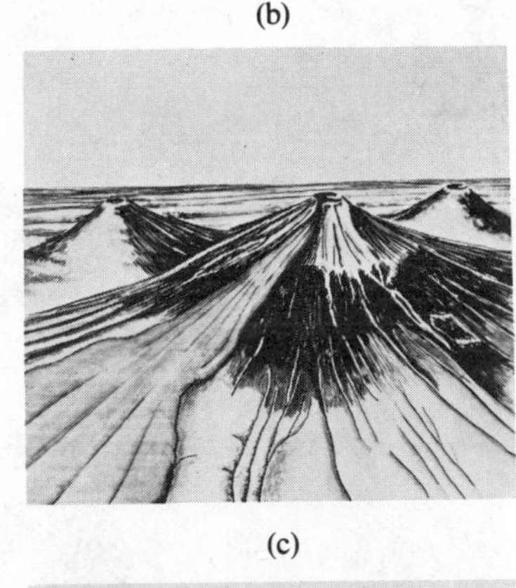

(c)

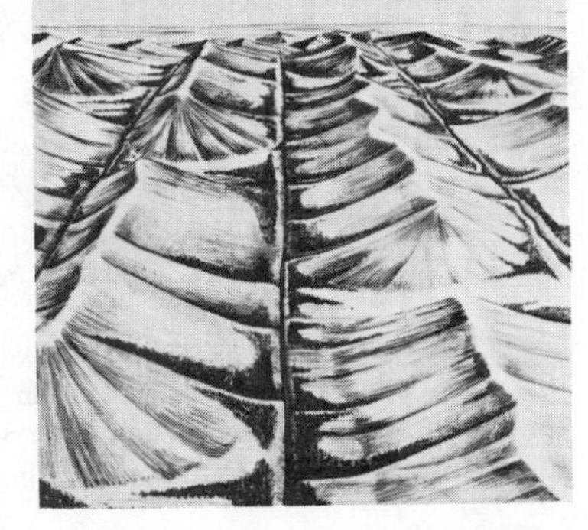

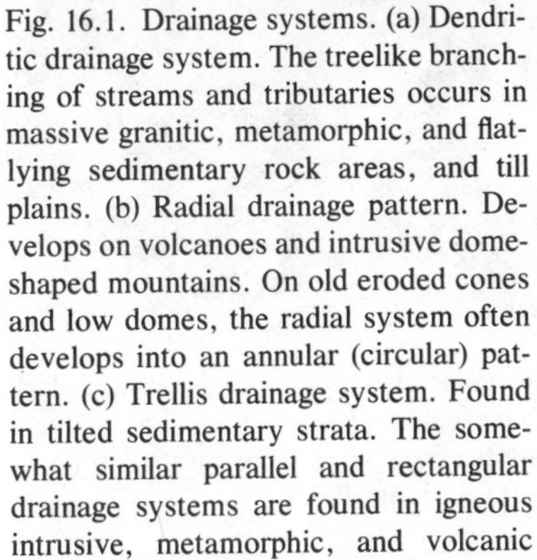

Fig. 16.1. Drainage systems. (a) Dendritic drainage system. The treelike branching of streams and tributaries occurs in massive granitic, metamorphic, and flat-lying sedimentary rock areas, and till plains. (b) Radial drainage pattern. Develops on volcanoes and intrusive dome-shaped mountains. On old eroded cones and low domes, the radial system often develops into an annular (circular) pattern. (c) Trellis drainage system. Found in tilted sedimentary strata. The somewhat similar parallel and rectangular drainage systems are found in igneous intrusive, metamorphic, and volcanic bedrock areas which contain well-developed faults and joints. Parallel drainage systems are also found in the sediments of lower coastal plains (U.S. Army 1963).

Special surficial markings

Important indicators of arctic and subarctic ground conditions are special surficial markings, such as patterned ground, solifluction features, and thermokarst depressions, that develop under the special climatic conditions that prevail there. Though most of the markings are visible to a person on the ground, certain ones are perceptible only from the air. The form of development of the markings is influenced by the vegetative cover, soil type, and slope, as well as by the climate. Markings tend to be fewer and less strongly developed in the subarctic, although some of the phenomena involved, such as ground thermal shrinkage cracking

Fig. 16.2. Button drainage system. In wide gullies having deep, fine-grained alluvial soils containing large masses of ground ice, a series of circular pools may form in the gully pattern as shown. During the breakup, large quantities of flowing meltwater induce substantial thawing (U.S. Army 1963).

under cold winter temperatures, are observed as far south as the temperate zones. In seasonal frost areas, however, the shrinkage cracks disappear each summer, whereas in permafrost areas the roots of the cracks are generally preserved in the permafrost throughout the summer, thus making possible the formation and growth of permanent ice wedges.

Patterned ground and terrain evaluation

General considerations on the formation of patterned ground as outlined in Chapter 5 may be amplified and extended to include in particular the evaluation of terrain for construction possibilities.

Field parties should keep in mind the following guidelines for construction in areas where patterned ground occurs:

(i) Try to locate roads and buildings as near a drainage divide as possible.

(ii) In location work, be wary of areas with scant lichen cover, as this may indicate significant movements in the active layer.
(iii) When building on sorted patterns, avoid disturbing the stones, which may provide subdrainage.
(iv) In selecting borrow, note now materials in the patterns are constituted and choose those which give the best grading for the purpose intended.
(v) Have equipment operators become acquainted with the various kinds of patterns and, by trial excavations, learn to correlate the properties of the patterns with the capabilities of the excavating equipment.

Frost creep, creep, and solifluction (gelifluction)

It is very important to detect any evidence of slope instability in the natural terrain. A structure, highway, railroad, pipeline, or other facility placed on a slope which experiences significant progressive downslope movement of soil or rock or which is subject to land slides may be damaged, made unserviceable, or even destroyed by such movements. Even if the movement is quite slow, such a facility as a highway or railroad may require periodic corrections of grade or alignment or reconstruction, and high maintenance costs may be experienced. Washburn (1967) has defined *frost creep* as the ratchetlike downslope movement of particles as the result of frost heaving and subsequent ground settling upon thawing, the heaving being predominantly normal to the slope and the settling more nearly vertical. Frost creep is thus fundamentally different from *creep* as conventionally defined in rheology, which is the continuing strain deformation of material under shear stress at rates so slow as to be usually imperceptible except by prolonged observation. *Solifluction* is the slow gravitational flowing from higher to lower ground of masses of soil sufficiently loose and saturated with water so that they behave like a viscous liquid. It is, in effect, a slow flow slide. Where frozen ground is involved it consists of thawed earth material flowing over a surface of frozen material. It may be presumed that when solifluction occurs in the arctic and subarctic, it is accompanied by frost creep. Washburn (1973) has pointed out that the term *solifluction* applies to any soil flow and is not restricted to cold climates; he has therefore proposed use of the term *gelifluction* for soil flow associated with frost action. The general term solifluction is the more commonly known term in soil mechanics, however, and is used here. Permafrost provides conditions especially favorable for solifluction, namely, an impervious base under

TABLE 16.1.
Forms of solifluction associated with frost action (gelifluction)

Form	Description
Solifluction sheet	An extensive, sheetlike mantle of soil moved by solifluction. It may exhibit striped forms of patterned ground and include a number of the types of solifluction landforms described below
Solifluction bench	A large, terracelike structure. The frontal scarp is quite steep—up to 60° from the horizontal—and may be from 100 to 4000 ft (30 to 1200 m) across and 2 to 20 ft (0.6 to 6 m) high. These benches occur on slopes of 5 to 15° or more
	Solifluction benches with a generally straight front tend to move slower than those with a strongly lobate front, which may move some inches in a year, and they often become stabilized and later break down through frost action. A lobate solifluction bench is therefore likely to be still active, and the area is less favorable for construction than is a solifluction bench with a straight front. Lobate benches grade into solifluction lobes
Solifluction lobe	An individual tongue of soil moving downhill on a steep slope of about 15 to 25°. The escarpment is from 1 to 5 ft (0.3 to 1.5 m) high and 10 to 30 ft (3 to 9 m) wide. The downslope length of a lobe is greater than its width and may vary from 20 to 150 ft (6 to 45 m). These lobes occur where there is sufficient water seepage to make the soil very fluid. They commonly move faster than the benches and indicate highly unstable conditions. Lobes grade into solifluction streams
Solifluction stream	A downslope-oriented streamlike area of solifluction with a length many times its width. Solifluction streams may be expected to have very unstable conditions within the ''channel'' area

Source: U.S. Army 1973.

overlying saturated thawed material throughout the thawing season. Ice concentrations at the interface between frozen and unfrozen soils may also cause this base to be slippery. Table 16.1 summarizes the characteristics of various types of solifluction features.

Frost sloughs, described by Lane (1948), are shallow slides that occur when the stability of frost-loosened and moisture-saturated fine-grained soils on slopes is reduced during thaw. Because of the relatively shallow depth of the thawed material over frozen soil, the slide dimensions are often restricted, and individual slides may often involve only a limited part of the slope area. In some cases the material may move as little as 1 to 2 ft before coming to rest with its surface vegetation still relatively intact, though deformed. In other cases the slide may continue as far as the bottom of the slope.

More deep-seated conventional slides, with the mass of soil tending to move downward and outward as a body under the influence of gravity, tend to be restricted in permafrost areas because of the strength of the frozen materials that must be involved.

Flow slides are slides in which thawed soil passes into a stage of complete liquefaction once failure starts, and the material flows rapidly as if it were a liquid, not stopping until the slope of the surface becomes quite small. These slides are very similar to slides of the same type occurring in temperate regions. Their scars are often visible long after occurrence of the flow and they are good indicators of unstable conditions. These are sometimes called *mudflows*.

Thermokarst

Thaw lakes, as described in Chapter 10, are lakes which have had their origins in the thawing of ground ice or have been enlarged thereby. They may typically have scalloped shorelines and almost vertical banks with overhanging vegetation and, in forest areas, leaning trees. *Oriented lakes,* a special form of thaw lakes, occur in the Alaskan Arctic Coastal Plain and have been also observed in the Mackenzie Delta area. The lakes are as much as 4.8 km in length, are roughly elliptical, and lie nearly parallel to one another, with their long axes oriented a few degrees west of north. Many are invaded by vegetation and may show polygons arranged in a series of concentric rings. Associated with the lakes are ridges of fine sand, as much as 1.6 km long, running approximately at right angles to the lakes. Other forms of thermokarst are *linear and polygonal troughs* formed where the ice-wedges thaw, *collapsed pingoes,* and *alasses*.

Vegetation

When used in conjunction with other information, vegetation may provide a useful guide to subsurface conditions, provided a correlation is established, verified, or known specifically for the area in which the investigation is being conducted. A ccrrelation which is valid for one

TABLE 16.2.

Characteristics of common subarctic trees

(arranged in order of increasing moisture demand)

Tree	Moisture and soil	Shade [a]	Temperature	Root system	Seed-bed favored	Main associates	Common [b] minimum depth to permafrost	Common locations	Remarks
				Aspen to Pines					
Quaking aspen	Moist to dry sandy silt. Never in waterlogged ground	Very intolerant	Warm	Deep semi-taproot	Bare mineral soil	Poplar, spruces	6 ft (1.8 m)	South-facing slopes; lakeshores; recent burned-over areas	Can exist in very dry locations
Balsam poplar	Moist sandy silts	Very intolerant	Fairly warm	Deep semi-taproot, strong laterals	Bare mineral soil	Chiefly pure	6 ft (1.8 m)	Riverbanks	Forming characteristic pure stands on recent alluvium
Alaska birch	Wide range of conditions	Intolerant	Cold	Broad and shallow	Bare mineral soil	White spruce	3 ft (0.9 m) (usually)	Sloping woods	Dwarf birches in tundra, etc., belong to different species
White spruce and pines	Best on silty soils, loess	Tolerant	Cold	Rigidly shallow, sometimes less than 1 ft (0.3 m) deep	Mineral or organic materials	Birches, aspen, poplar	3 ft (0.9 m) (usually for spruce)	Woods or good soils	Other conifers occur. Pines chiefly in drier areas, scrubby around muskegs
				Willow to Tamarack					
Felt-leaf willow	Damp silty to gravelly soils	Intolerant	Cold	Variable	Bare mineral soil	Other willows, or pure	3 ft (0.9 m) (usually)	Natural levees, marshes, lakesides	Widely distributed as tree or shrub

Mountain alder	Damp to wet silty organic soils	Fairly tolerant	Cold	Shallow	Litter and humus	Various or pure	3 ft (0.9 m) (usually)	Lake margins and protected gullies; understory to birches and white spruce	Indicative of seepage and of small watercourses in tundra
Black spruce	Swampy soils preferred	Fairly tolerant	Cold	Shallow, many windfalls	Chiefly bare mineral soil	Tamarack and muskeg vegetation, pure open forest in north	3 ft (0.9 m) (usually)	Swamps, stony slopes, sometimes as a forest type	Also found in dry situations
Tamarack (larch)	Swampy soils, waterlogged areas	Very intolerant	Cold	Broad and shallow	Litter and humus	Black spruce and muskeg vegetation	3 ft (0.9 m) (usually)	Swamps, usually as isolated trees; riverside flats	Occasionally found in drier situations, especially in north

[a] Tolerance to shade is largely relative, often variable, and consequently difficult to judge as well as uncertain.
[b] Figures below these common minima are not rare for some trees.
Source: (U.S. Army 1963)

TABLE 16.3.
Tundra vegetation groups with reference to soil moisture

Type	Location	Ground and vegetative cover	Patterned ground	Estimated annual frost zone	Remarks
Dry	High, dry locations (terraces, etc.)	Few plants, but usually scattered grasses and other small herbs and ground shrubs—with lichens on bare earth between them	Unusual	3 to 10 ft (0.9 to 3 m)	Usually good construction sites. Indicators include crowberry, alpine holygrass, clumps of purple saxifrage
Moist	Gently rolling to flat alluvial and glacial deposits	Except in the far north, vegetation commonly forms a complete mat, the plants being up to 1 ft (0.3 m) (occasionally up to 3 ft [0.9 m]) high. Heath shrubs, sedges, grasses, mosses, and lichens	Common	18 in. (0.45 m)	Bushy willows, alders, and scrub birches may be found in protected ravines—usually fair construction sites
Wet	Depressions and obliterated lakes	Vegetation underlain by peat often forms a thick mat with peat mounds and ridges up to 4 ft (1.2 m) in height. Sedges, tall grasses, tussock grasses, blueberries, and mosses all in evidence	Common	1 to 4 ft (0.3 to 1.2 m) (wet to drier)	Generally less extensive than the other types. Pingos may occur with wet tundra—poor construction sites. Indicators include tall sedges and coarse tussock and other type cottongrasses

Source: (U.S. Army 1963)

area may not necessarily be valid in another geographically or environmentally different area. Factors such as amount of sunshine, topographic position, soil moisture, summer soil and air temperatures, and exposure may affect plant growth, as well as permafrost levels or soil texture, and their effects may combine in many different ways to affect plant growth differently at different locations. Plants have certain basic requirements for their survival and growth, however, and simply by knowing the basic requirements of the plants growing at a particular location, one can infer the soil and permafrost conditions.

Vegetation types are often more easily identified from low-altitude, oblique aerial photographs than from vertical ones, which require considerable interpretive experience. The Corps of Engineers has prepared a comprehensive review of tree species and their airphoto patterns in Alaska, which is presented in the US Army Manual (1963) *Terrain Evaluation in Arctic and Subarctic Regions*. This information should be used with care in other areas than Alaska, as growth habits of trees of identical or closely similar species may be different in those areas.

For the subarctic, where the summers are long enough and warm enough to permit trees to grow, Table 16.2 presents a summary of the characteristics of a number of trees usually found in the subarctic areas of Alaska and northwestern Canada, in terms of their growth requirements. Table 16.3 summarizes vegetative cover and other characteristics of tundra terrain for dry, moist, and wet soil moisture conditions.

17. Route or site selection and development

Introduction

Because of the relative lack of development of the arctic and subarctic, the selection and investigation of a site or route may involve virtually unexplored terrain for which there is little or no existing soil information. A wide opportunity for choosing a site or route may then be present, and major savings in construction and maintenance costs may be possible from a well-conducted program of investigation, and the risk of engineering failure may be greatly reduced. Even if the site or route is located within a relatively developed area where some general soil information exists and the room for choice is much more limited, significant advantages may still be possible from even limited adjustments of alignment or siting. The specific information to be acquired will vary with the site selection or development problem, and judgment is required in formulating a good program of investigation and analysis. The procedures outlined in this chapter are applicable to a major project, but they can be easily modified or reduced in scope as necessary to suit different situations and needs. Through proper attention to subsurface conditions during the site selection, soil mechanics design and construction problems can be simplified.

A number of the kinds of data shown in Fig. 13.1 and mentioned in following paragraphs may at first glance appear unrelated to the soil investigations. Nevertheless, to prepare a practical soil mechanics design, more than purely soil-related information is needed. For example, the choice of a foundation type, a road cross section, or embankment materials at a remote location may depend on the feasibility and relative costs of transporting various types of construction equipment to the location or on the relative availability of processed coarse aggregate or other special materials.

Preliminary studies

The first step in the investigation should be to collect available information on the area in which the site or route is to be located. This area may involve in some cases thousands or even tens of thousands of square miles. Reports, maps, airphotos, and climatic and other data should be obtained from governmental agencies, such scientific and technical groups as geographical societies, libraries, engineering and commercial firms, photointerpretation and remote sensing specialists, travel reports, scientific and technical papers, exploration records, reports of persons who may have done work in the area, and any other available sources. Geologic and topographic survey maps, aeronautical charts, and maps prepared from airphotos may be used. Aerial photographs may be available in stereo pairs. Photographic or other information obtained from high-altitude aircraft or satellites may be available. If suitable photography is available, it may be possible to prepare a vicinity map specially for the project. Aerial photographic coverage of some type is now available for most northern regions in North America. If no maps or aerial photographs are available, maps suitable for the preliminary studies can be prepared by obtaining high-altitude, wide-angle, small-scale aerial photographs. Agricultural or pedological survey maps may sometimes be available. Interviews with those who have personal knowledge about the area may be helpful. Sometimes well drilling logs or even foundation exploration records for other construction within the area may be available.

Reconnaissance

Reconnaissance studies are made for the purpose of determining the most feasible general site areas or route corridors for detailed investigation. This may be accomplished by a combination of aerial and ground reconnaissance.

Air reconnaissance

Reconnaissance flights by fixed-wing aircraft or helicopters or both, in combination with aerial photography and, if appropriate, airborne geophysical exploration for presence of massive ground-ice, are especially valuable in initial studies to obtain data on such factors as general terrain conditions, surface icings as revealed at the end of winter, flooding conditions, the sequence of snow melt during spring thaw, possible water or land landing areas for ground reconnaissance parties, and possible survey and construction camp locations.

Terrain analysis from aerial photographs is especially useful in studying remote areas for which there is little detailed information. Aerial photographs can be used to identify potentially suitable sites or routes and also areas to be avoided. They can further be used to delineate the boundaries of the various soil types present at the ground surface and to predict their engineering characteristics. Limits of frozen and unfrozen soils can also be estimated. An expert analyst of aerial photographs can identify surface formations suitable for construction sites or highways with high accuracy. Interpretation of aerial photographs is not suitable for predicting conditions at depths below the upper few feet of the ground. Frost, McLerran, and Leighty (1966) have recommended scales of 1 to 40 000 to 1 to 100 000 for regional photointerpretation studies. Mollard and Pihlainen (1966) have recommended a scale of 1 to 40 000 for area studies.

Ground reconnaissance

Ground reconnaissance is used to check the information obtained from preliminary studies, air reconnaissance, and aerial photographs and to obtain additional reconnaissance-type information not otherwise available. Observations should be made concerning weather, geology, topography, permafrost, soil, rock outcrops, stability of slopes, snow cover, icings, vegetation, groundwater, surface water, flooding and drainage conditions, sources of water and materials for construction, and any other pertinent factors. In the case of road, pipeline, or surface communication lines, spot investigations of key points along potential routes may be made. Local residents, if any, should be queried for information they may have. Shallow explorations may be made by hand shovels or hand augers if the ground is thawed, topographic information may be obtained with hand levels, and a complete photographic record should be made to assist the subsequent office studies.

Locations of any existing installations, roads and communication facilities, wells, borings, test piles, and natural exposures, availability of local labor, construction equipment and materials, and means of access for both survey and construction teams may be determined in relation to the possible site or route locations.

Localization of tentative sites or routes

Based on the information gathered in the preliminary and reconnaissance studies, a limited number of the most promising areas or corridors for the proposed construction should next be selected for somewhat closer examination. These studies may be carried out using principally airphoto interpretation, supplemented by additional ground checks, to obtain new information or to verify predictions, and airborne or ground geophysical investigations if needed. New photography concentrating on the better sites or routes should be obtained if satisfactory photography is not already available. Points to consider for photography at this stage, as recommended by the U.S. Army (1966b) for site selection investigations by airphoto interpretation in an unmapped area, are as follows:

(1) Types of photographs. Nine- by nine-inch vertical and oblique photographs are most commonly used (9- by 18-inch [229- x 457-mm] photographs are cumbersome for field use). Low-altitude obliques are useful for evaluation studies and illustrative purposes. Stereo-pairs greatly facilitate terrain interpretation.

(2) Focal length of lens. A short focal length lens should be used in flat areas to increase the apparent depth perception in the stereoscopic image so that minute changes in relief are resolved. A 6-inch (approx. 150-mm) lens is recommended. In hilly or mountainous terrain a 12-inch (approx. 300-mm) focal length lens is most practical.

(3) Type of film. Panchromatic film is widely used. Occasionally, strips of infrared or Kodacolor aero are of value for interpretation purposes.

(4) Type of filter. Filters are used to cut atmospheric haze and to accentuate tonal differences. Yellow haze filters, often referred to as "minus blue" filters are used with both panchromatic and infrared films. If light conditions permit, a medium red filter "Wratten No. 25" is preferable.

(5) Overlap and sidelap. Photography intended for preparation of mosaics and for detailed interpretation purposes should have a 60 percent overlap and a 30 percent sidelap. For trimetrogon photography, the interval of flight lines is specified by the photo interpreter.

(6) Location. Geographical coordinates bounding the area should be indicated. Flight lines plotted on trimetrogon obliques or large-scale topographic maps are of considerable value to the aerial photographer. If possible, checkpoints should be established on the ground to aid the aerial photographer.

(7) Scale. The scale of photography is normally specified in terms of the representative fraction (RF) which is equal to the flying height (in feet) of the aircraft above mean terrain divided by the focal length (in feet) of the aerial camera. For example, the scale of photography flown at 6,000 feet (1829 m) with a 6-inch (152-mm) lens is 1:12,000.

(8) Season for photography. Winter photography is seldom desired because snow masks surface features. Usually the summer season is specified because shadows are shorter, but photography taken during the breakup period, when deciduous trees are without foliage, is generally most useful.

(9) Types of prints. For most interpretation work either glossy or semimatt prints are acceptable. Double-weight paper for field use is advisable.

(10) Mosaics. Uncontrolled (distorted scale) mosaics or index sheets are of value in the localization stage. Controlled (true scale) mosaics are desirable in the stage following localization.

Using the collected information and the aerial photographs, one can then prepare rough engineering soil maps for the areas or corridors under consideration. Additional ground reconnaissance may be made if necessary. Based on a comparative evaluation of the merits of the potential site or route locations as developed on these maps, a single general site or route corridor may then be chosen for detailed studies.

Detailed investigation of the best general site or corridor—initial phase

When a single general site or corridor has been chosen as the best area for the proposed project, additional air photography may be obtained at a scale of 1:4800 for a construction site or 1:12 000 for an access road, highway, pipeline, or utility line route using the same camera and lens combination as in the previous studies. Photography may typically cover about 5 miles on either side of a tentative route alignment. Ground control should be established if possible. One or more engineering soil maps, as necessary, may then be prepared, using these large-scale photographs, on which information may be shown in more detail than on the previous rough maps. Additional field data may be obtained if needed, to aid preparation of these maps. The types of surface cover should be mapped; this information will not only help in interpretation of subsurface conditions but will also aid survey crews, delineate stands of usable timber, and aid in development of insect control and environmental protection plans.

The initial phase of the detailed investigations is most effectively carried out in permafrost areas when the depth of summer thaw has reached or nearly reached its maximum, between the latter part of the summer and the freeze-up period. Parties may then go into the field with copies of the aerial photographs, the surface cover maps, and the engineering soil maps. Areas of distinctive aerial photographic patterns which should be investigated and tentative locations of initial subsurface exploration borings should be marked on these documents. One or more potential locations for the planned facilities or route alignments within the study boundaries may be marked on the documents.

In this initial phase of the detailed investigations, the objective should be to expand the information on the site or route sufficiently to permit a decision on a specific facility location or layout or on a specific route alignment. Information to be obtained or verified in this stage of the investigations should include:

Preliminary location and topographic ground survey data.
Drainage, groundwater, and water supply information.
Tree, plant cover, and burned-over area details.
Special terrain details (polygons, earth slides, etc.).
Horizontal limits and continuity of soil formations.
Soil and rock conditions with depth.
Seasonal frost and permafrost conditions.
Ground temperatures.
Weather conditions, especially air temperatures.
Sources of aggregate and borrow materials.

From the paper locations, the tentative facility positions or route alignments may be staked out in the field, and approximate line and grade and topographic information can be obtained.

Drainage, groundwater, and water supply information should include data on lakes, rivers, and streams, icings, erosion, seepage emergence points, springs, high water marks, and depth to water table. When temperatures drop below freezing, measurement of groundwater levels in observation wells becomes difficult because the water in the observation pipes freezes. Plant cover, including trees, brush, and low ground cover, should be identified as to species, size, and density. In areas which have experienced forest fires or other disturbance of the surface cover, the effects on the permafrost table and the thermal regime should be specially measured and the year of disturbance determined if possible. Special surface features to be identified include polygons, various forms of patterned ground, slides, solifluction, thermokarst depressions, and other frost-related formations. The horizontal limits and continuity of soil formations should be determined by explorations to a minimum depth of 10 ft (3 m) or to non-frost-susceptible granular soil or to bedrock. Each change of soil conditions with depth should be carefully recorded in a field log of each exploration, with soils, ice, and rock classified visually. Boundaries between frozen and thawed layers should be identified in both the active layer and permafrost strata. Surface geophysical surveys may be used under favorable conditions to explore subsurface conditions where substantial areas have to be covered.

At least two thermocouple assemblies should be installed, one under a stripped surface, one under natural cover, with sensors located at 1-ft (0.3 m) intervals from immediately below the ground surface to a depth of at least 10 ft (3 m) and preferably to about 30 ft (9 m); readings may be made weekly. A simple weather observation station for air temperature and other basic measurements should be established. In some areas, the search for sources of suitable borrow materials, for special purposes such as road base, portland cement concrete, or slope protection, may require a major, separate investigation. In some locations such materials may have to be transported from sources tens or even hundreds of miles away, and the seasonal feasibility of the required water or land transportation may be a key controlling factor in establishing the overall project schedule.

Detailed investigation—final phase

Once proposed final paper locations or alignments have been determined on the basis of the results from the initial phase of the detailed investigations, they can be staked out in the field, tentative final lines and grades and topographical information can be obtained, and detailed subsurface investigations can be initiated. At this stage, borings may be drilled to whatever depths are needed to obtain necessary design information for planned facilities and to determine more fully the continuity and extent of strata.

Special care is required when bedrock is encountered in the investigations. Extra drilling may be required to distinguish between bedrock and boulders or large pieces of free rock when encountered in borings. Again, the exploration team should be aware that, in seasonal frost areas, severe frost heave frequently occurs in bedrock because of mud seams in the rock or concentrations of fines at or near the rock surface, together with the capacity of fissures in unfrozen strata to supply ample quantities of water for growth of ice lenses. In permafrost areas, experience shows that permanently frozen bedrock may contain substantial ice inclusions, especially in the upper 15 ft (4.5 m), which can produce severe settlement if allowed to thaw. At Thule, Greenland large amounts of ice were discovered to substantial depth in shale bedrock at the crest of a high ridge with excellent surface drainage conditions. Undisturbed frozen cores should therefore be obtained from at least the upper 10 to 15 ft (3 to 4.5 m) of bedrock in permafrost and should be visually examined for ice inclusions whenever the design or performance of the proposed facility could be significantly affected by the presence of such ice inclusions.

If a thermocouple or thermistor ground temperature assembly was not installed to about 30 ft (approx. 9 m) deep during the initial phase of the detailed studies, one should be installed to that depth or preferably deeper at this point in the investigation. Exploratory well drilling for a water supply can also be included in the program, if applicable. For design purposes, pile installation, pile load, plate bearing, anchor pull-out, or thaw-consolidation field tests may be performed as necessary.

Techniques for direct subsurface exploration

The extent of direct site exploration should be sufficient to reveal the extent of frozen strata, permafrost and excess ice, including ice wedges, moisture contents and groundwater, and the characteristics and properties of frozen and unfrozen materials. In permafrost areas, the active layer is relatively easy to penetrate during the thaw season, and hand augers or hand-dug test pits may be used in thawed soils above the level of any water table; difficulties may be encountered in keeping test pits or uncased drill holes open and in soil sampling, because of the high water table in the active layer during this season. For this reason, and because access to drilling sites with heavy equipment and operations on the surface of all kinds are easier in the winter in many locations, it is frequently preferable to carry out explorations during the colder parts of the year when the active layer is frozen. Because frozen soil may have the strength properties of lean concrete, hand-exploration methods are not practical in frozen materials. On exploration projects in locations where access is difficult or limited, lightweight drill rigs, designed so that they can be disassembled into individual pieces small enough to be carried in light aircraft, have been used. Conventional core-drilling techniques using above-freezing drilling fluids do not produce successful results in materials containing excess ice because the ice is melted. Core drilling with drilling fluid refrigerated to temperatures below freezing permits the recovery of cores from even bouldery soils and bedrock which are almost completely undisturbed. These cores allow the thicknesses of any ice layers present to be measured manually and allow testing for in-place properties (Hvorslev and Goode 1966; Lange 1973*a*). Drive-sampling is feasible in frozen fine-grained soils at temperatures down to about −4°C and is often considerably simpler, cheaper, and faster (Kitze 1956). Samples of frozen soils obtained by drive-sampling are partially disturbed but yield accurate moisture content values. Test pits excavated by bulldozer or other power equipment may offer an effective solution where such equipment is available. Truck-mounted power augers employing tungsten carbide cutting teeth are quick and efficient in frozen soils which do not contain too many cobbles and boulders, if only classification, gradation, and approximate ice content information is needed. Large track-mounted power augers are capable of drilling large diameter holes to as deep as 25 m; these holes can be entered, logged in detail, and sampled so long as the material is safely ice-bonded. Heavy, track-mounted rigs have also been used with oil-well-type rotary bits and compressed air removal of cuttings, permitting rough ice content determinations. Such equipment is far too large and heavy for most subsurface investigation projects, however.

In summer, the depth to permafrost can often be determined most quickly by probing. For soils offering considerable resistance or where a frozen layer must be penetrated regular power-operated drill rigs of various types may be used. A pneumatic drill may also be used. In relatively soft, fine-grained soils, hand-probing is often effective; a sharpened steel rod may be pushed to depths of as much as 15 ft (4.5 m) using clamp-grip pliers. After enlarging the hole by some method, additional depth may be gained. Test pit and auger explorations to permafrost may also be used.

Borehole permeability or pumping tests may be performed if desired when unfrozen materials are being drilled.

Field logs are prepared by the exploration foreman or inspector as exploration proceeds. Samples for routine laboratory analysis are usually recovered at each change of material or at 5-ft (1.5- m) intervals. Continuous sampling is performed where a continuous record of soil conditions with depth is needed. Materials are field-classified, and samples are retained, labeled as to exploration number, sample number, depth, and date, and forwarded to the laboratory for verification of the classification. When rock is penetrated, the log is continued in the same manner as for soil, and the rock is classified in accordance with accepted geologic designations. Disturbed soil samples intended for identification and moisture content tests may be placed in moisture-tight, screw-top containers. Samples intended for compaction or aggregate tests are placed in cloth or plastic bags. Rock cores are usually filed in sequence in specially constructed, wooden core boxes. Undisturbed samples of unfrozen soils may be preserved by techniques described by Hvorslev (1949) in his important publication on subsurface exploration and sampling of soils. Samples to be kept frozen must be placed in insulated containers until they can be delivered to a laboratory cold room for examination and tests. Refrigeration of clay-type materials or soils containing saline pore water at temperatures which differ from in-place values may change ice contents. On a large or especially important investigation, a laboratory cold room may be established in the field.

In the laboratory, a laboratory log is prepared as shown in Fig. 17.1 (George 1973). This is a copy of the initial and final sections in the log of an actual boring at a dam site in Alaska. Note particularly the essential information concerning boring number, location, elevation, and type of drilling equipment used, which is recorded at the top of the log sheet, the solid bar (on the left) designating frozen materials; the number of blows per foot (recorded on the right) is a measure of the resistance or 'relative density' of the strata.

Geophysical surveys

Geophysical surveys permit relatively large areas or long distances to be explored at a low per-unit-area cost, in minimum time. In permafrost areas they offer a potential for mapping high ice-content areas and for delineating areas of permafrost in discontinuous permafrost areas. Freezing of water in soil and rock causes changes in physical properties that can be measured by various procedures, either from the air or on the ground. State-of-the-art reviews of geophysical methods that have been considered for use in permafrost areas are contained in papers by Barnes (1966), Ferrians and Hobson (1973), Linell and Johnston (1973), and, most recently, by Scott, Sellmann and Hunter (1979). Both resistivity and refraction types of conventional seismic systems have been found useful in permafrost areas (Garg 1973 and Hunter 1973). Seismic refraction can provide information on the depth to the upper surface of permafrost and on the physical properties of the permafrost through the seismic velocities. It can be used only under conditions of increasing velocities with depth, however. Because of the very large change in electrical resistivity that occurs when soil freezes, the electrical resistivity method can also be very useful under favorable conditions, although

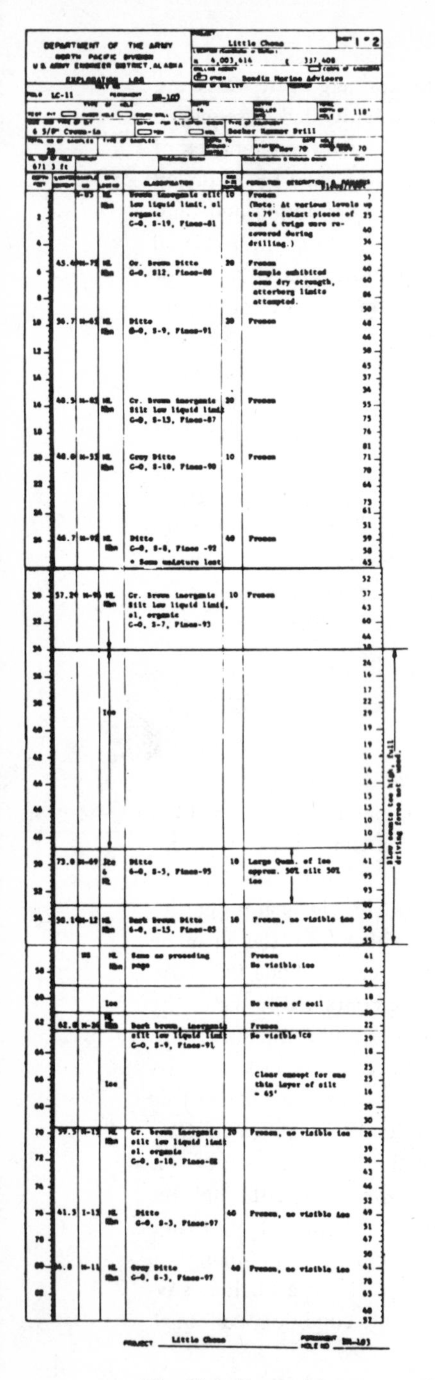

DEPARTMENT OF THE ARMY
NORTH PACIFIC DIVISION
U.S. ARMY ENGINEER DISTRICT, ALASKA
EXPLORATION LOG

Project: Little Chena — Sheet 2 of 2
N 4,003,614 E 337,408
Bendix Marine Advisors
Hole: LC-11 — Permanent: DH-103
Depth of hole: 118'
6 5/8" Crown-in — Becker Hammer Drill
Total no. of samples: 20
Started: 6 Nov 70

Thermocouple pipe set to 117' depth

Bottom of hole at 118'. Not to bedrock. Drill pipe was plugged at 118'.

*Believe bottom 100' intact but joint 17' below ground, surface some apart but was forced back together. Seal probably not adequate to hold or prevent seepage of fluids. This was pretty well verified when we later tried to fill pipe with oil.

**Last blow count for ½ of a foot.

Project: Little Chena — Permanent hole no.: DH-103

Fig. 17.1. Top and bottom sections of Little Chena Dam exploration log DH-103 (George 1973).

it can also pose problems in interpretation. Electro-magnetic systems, discussed in the last three of the review papers cited above, and particularly in the one by Scott, Sellmann and Hunter, have been found capable of successfully distinguishing materials such as soil, rock, and ice with depth; these systems are in a state of rapid, continuing development, and anyone contemplating their use should first check the state of the art. Development of techniques for satellite acquisition of data or other sophisticated approaches may ultimately provide new methods for obtaining useful information on ground details.

Soil tests required

Depending on the size of the project and the type of soil information required, samples may be tested in a portable laboratory in the field or transported back to a permanent laboratory, or a combination of these approaches may be used. Test requirements for various purposes have been discussed in preceding chapters, the most basic being the description and classification tests. Density, specific gravity, moisture content, permeability, compaction, frost susceptibility, thaw-consolidation, swell potential, or other tests may be performed as appropriate, and results may be included on the laboratory log. Procedures for tests are discussed in Appendix A. Special field tests, such as pile loading tests of piles in permafrost (discussed in Chapter 13), may require special procedures.

Frost and permafrost observations

As the boring work progresses and thermal conditions around the ground temperature installations stabilize, the ground thermal regime will begin to be revealed. The depth of the active layer and the position of the permafrost table at the end of summer or beginning of fall should be recorded. Subsurface temperature observations made from late summer into fall should be frequent enough to determine the highest seasonal ground temperatures in the upper part of the permafrost because they are useful in foundation design. By continuing the observations through winter and spring into summer, the coldest extremes of ground temperature with depth can be obtained, and in seasonal frost areas the maximum seasonal depth of frost penetration can be measured. If air temperatures and ground surface temperatures for various surface conditions are observed through full thawing and freezing seasons, n-factors may be computed. By correlating air temperatures with the corresponding simultaneous air temperatures at one or more of the nearest long-term regular weather stations and using the statistical records of these stations, long-term average and extreme temperature, freeze and thaw index, n-factor and freeze-and-thaw penetration values can be estimated approximately. These data may in turn be used in evaluation of any degradation, thaw-settlement, and frost-heave effects observed at the location, as well as for engineering design.

Procedures for obtaining in-place measurements of subsurface temperature conditions during the design soil survey and for monitoring these conditions during and after construction are reviewed briefly in Appendix B.

Appendix A. Engineering soil testing

Soil tests commonly performed for engineering purposes are shown in Table A.1. In the upper part of the section of the table which is headed *Tests on Unfrozen Soils,* tests such as grain size analysis and liquid and plastic limit tests, do not measure clearly defined engineering properties. These tests must therefore be performed in strict accordance with standardized procedures if they are to convey well-understood and consistent meanings to the design engineer. Procedures for tests in the lower part of this first section of the table, such as consolidation or direct shear, which determine the engineering properties of soils, may be varied or adjusted to best meet the needs of the investigation, although basic and accepted procedures will normally be used as bases.

In the tests on unfrozen soils listed in Table A.1, standard handling, storage, and testing precautions, such as the use of a humid room for storing and processing undisturbed specimens, should be adhered to. These tests are well known to soil mechanics engineers, are well described in available references, and will not be discussed further here.

The tests listed in the second section of the table, for frozen conditions, include tests in which the materials remain frozen during the test, may become frozen during the test, or may be frozen only at the start of the test. Until they are tested, undisturbed frozen samples must be kept continuously below freezing at temperatures which will preserve the ice content and internal ice crystal structure as nearly as possible in its natural condition.

To obtain the dry unit weight and natural moisture content of a frozen core or lump specimen, the bulk density of the sample should be first obtained. The sample should then be thawed, and the dry weight of solids and the moisture content as a percentage of dry unit weight should be obtained. From these data, from measurements made of any ice layers visible in the specimen when frozen, and from knowledge of the approximate moisture contents these materials would have in-place if unfrozen, the existence of a settlement problem on thawing may be immediately apparent if the amounts of ice in excess of the normal void space are appreciable.

If the excess ice is frozen in interstitial form or is otherwise difficult to measure or if quantitative settlement predictions are needed, thaw-

TABLE A.1.
Engineering soil tests

Test	References for test procedures				
	AASHTO 1978	ASTM 1979	Lambe 1951	Corps Engrs 1970	Other
Tests on unfrozen soils					
Grain size analysis	x	x	x	x	
Liquid limit	x	x	x	x	
Plastic limit and plasticity index	x	x	x	x	
Specific gravity	x	x	x	x	
Organic content	x				
Moisture content		x		x	
Dry unit weight (lab.)				x	
Dry unit weight (field)	x	x			
Shrinkage factors	x	x		x	
Compaction	x	x	x	x	
Permeability		x	x	x	
Consolidation		x	x	x	
Direct shear	x	x	x	x	
Triaxial compression		x	x	x	
Unconfined compression	x	x	x	x	
Modulus of soil reaction		x			
Calif. bearing ratio	x	x			

Pile load test	x	
Resilient modulus		Thompson and Robnett 1979
Tests involving a frozen condition		
Moisture content		See text
Dry unit weight		See text
Thaw consolidation		Crory 1973
Direct shear		Frost Effects Laboratory 1952 Kaplar 1953, 1954 Roggensack and Morgenstern 1978
Triaxial compression		Sayles 1973
Unconfined compression		Sayles 1968
Frost susceptibility		Kaplar 1974 Penner and Ueda 1978
Pile load test		Linell, Lobacz, et al. 1980

consolidation tests may be performed on frozen specimens using procedures described by Crory (1973) and others. Standard consolidation test apparatus may be used, with a nominal initial load applied prior to thawing.

Direct shear, triaxial, and unconfined compression tests may also need to be performed on frozen specimens. As explained in Chapter 8, the tendency for frozen soils to deform progressively in creep requires special test and analysis procedures. Procedures for direct shear tests on frozen materials are described in reports by Frost Effects Laboratory (1952), Kaplar (1953, 1954), and Roggensack and Morgenstern (1978). Procedures for unconfined compression and triaxial tests on frozen soils have been described by Sayles (1968, 1973).

Standardized frost susceptibility test procedures, based on measurements of rate of heave, have been described by Kaplar (1974). These tests, which are run in an open system with unlimited availability of water at the base of the freezing specimen, provide a relative measure of frost susceptibility. Experience has shown that soils with average heave rates of 1 mm per day or less in this test (see the scale at left-hand side of Fig. 8.1) usually show acceptable performance under pavements. Using this procedure, the Corps of Engineers has evaluated about 250 soils from sites in cold regions around the world. Penner and Ueda (1978) have also proposed a soil frost susceptibility test in which a step-change freezing temperature is imposed at one end of the specimen and the heave rate is measured.

Appendix B. Monitoring soil and foundation behavior

Introduction

Soil and foundation behavior in areas of both deep seasonal frost and permafrost should be carefully observed visually and with instrumentation, both during construction and during the subsequent operational life of a facility. If severe problems are occurring, evidences of frost heave, settlement, slope instability, or detrimental surface and subsurface moisture movements may be visually apparent from soil contact marks on the foundation, changes of grade and plumbness in structures, displacements, cracks in walls, breaks in pipes, ponding and changes in direction of surface drainage, increasingly wet ground conditions, soil erosion and mass movements, and other phenomena. Often problems may develop slowly, and years may pass before difficulties become evident. If problems are developing, instrumentation is needed to give timely warning of the need for corrective measures and to provide the hard data needed for design revisions.

Inspections

Periodic inspections should be made, both during and after any construction or changes in the terrain, for the specific purpose of detecting any developing problems. Engineering works constructed in arctic and subarctic areas are usually quite expensive, and it is only prudent to take this elementary action to safeguard the project investment. It is not always easy to ensure that such measures will be taken. The engineers responsible for the design and construction of the facility will usually go to other assignments as soon as their jobs are complete, and those who become responsible for the subsequent operation and maintenance of the facility in following years may have inadequate understanding of the problems which may arise. Even when a detailed operation and maintenance manual covering these questions is prepared by the design engineers, its significance may not be conveyed through successive changes in the management of the facility, and experience shows that the manual itself may soon become lost. The only apparent remedy for this situation appears to be continuing, repetitive, and unremitting emphasis by higher authority and concerned agencies upon the need for periodic continuing checks of the performance and stability of facilities constructed under adverse terrain conditions.

Vertical movements

A system of measurement points and a frost-stable benchmark installed during construction will permit level readings to be made at intervals to detect any frost heave, thaw-settlement or other vertical movements, particularly differential movements, which are most structurally damaging. A design for a frost-stable benchmark is shown in Fig. B.1. By plotting the level readings against time, one can perceive the trend of any movements.

Ground temperatures

Thermistor or copper–constantan thermocouple assemblies installed in the ground prior to and during construction will permit monitoring of thermal regime changes during construction and during the life of the facility. Any increase in the depth of the active layer or any formation of a residual thaw layer will be detectable, as well as any tendency of permafrost temperatures to warm above the levels assumed for the design. If a ventilated foundation is not achieving required freezeback during winter or if thaw is reaching unsafe depths around foundation piles in permafrost, the ground temperature records will show it if the temperature assemblies have been well-located.

Thermocouples are relatively simple and cheap to use, but under winter conditions problems are encountered in keeping reference temperature ice baths, standard cells, and potentiometers at temperatures which will ensure accurate results. Thermistors of a select type, stabilized and properly calibrated, offer greater precision than is possible with thermocouples, but a heated, shock-protective enclosure is still required for the Wheatstone bridge under winter conditions, and other problems are possible. With either thermocouples or thermistors, painstaking care and awareness of potential pitfalls are required to ensure accurate, reliable data.

Various other systems for monitoring depths of freeze or thaw exist, such as inserting into the ground tubes filled with moist sand to which has been added a dye that changes color upon freezing of the moisture, but none provide the same combination of detail in data and reliability,

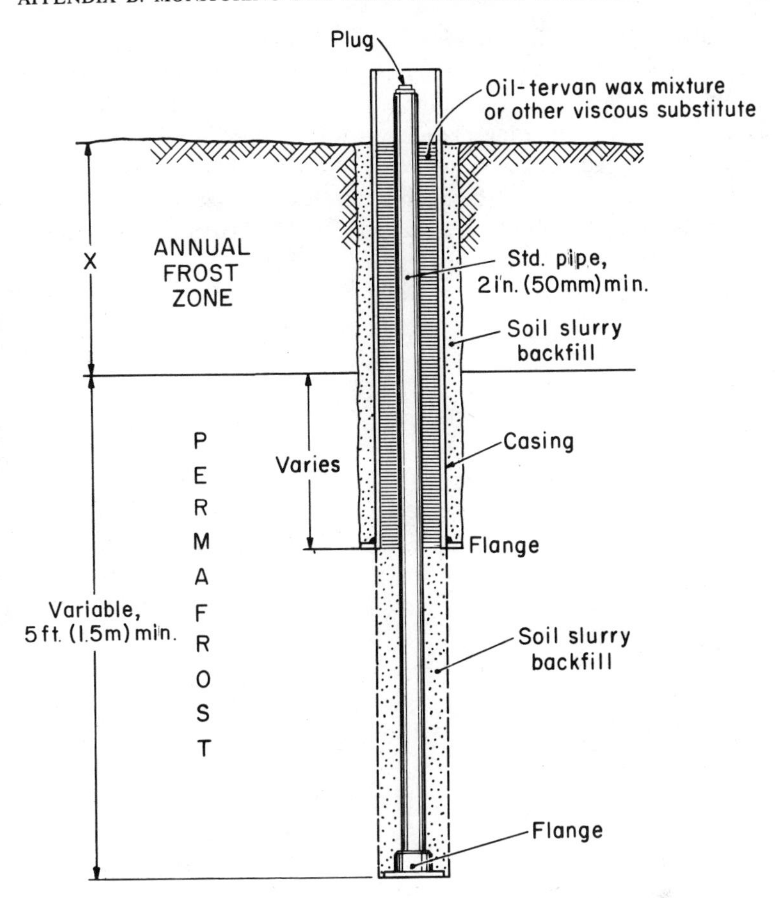

Fig. B.1. Recommended permanent benchmark (Linell, Lobacz, et al. 1980).

particularly for use under structures or other relatively inaccessible locations, as is provided by thermocouples and thermistors.

Monitoring groundwater

Sometimes the need may arise to monitor groundwater levels in the active layer overlying permafrost or in areas of deep seasonal frost penetration. Simple pipe observation wells are effective for this purpose in summer, but difficulties are encountered in fall and winter because water standing in the well freezes at the top when the frost line reaches it. If the groundwater level is dropping, the water within the well will be held up in contact with the ice until the latter is cut through. The water level will then drop, but time must be allowed for equilibrium to be established before valid readings can be obtained. No readings will be possible, of course, when soil above permafrost freezes completely. Observation wells through permafrost will freeze solid within the permafrost stratum unless there is water flow out of the well.

Appendix C. Literature

Technical literature of the arctic is generally quite specialized and not always readily accessible—even in many of the larger libraries. It is necessary, however, to have access to and a working knowledge of certain specialized publications. The following list of publications should provide adequate background for initial investigations of arctic soils and permafrost.

Arctic and Alpine Research. Institute of Arctic and Alpine Research, University of Colorado, Boulder, Colorado 80309, U.S.A.

Arctic. Arctic Institute of North America, University Library Tower, 2920 24th Avenue NW, Calgary, Alberta, Canada T2N 1N4.

Arctic Bibliography. Prepared by the Arctic Institute of North America. U.S. Government Printing Office, Washington, D.C. Beginning with Volume 13, published by McGill–Queen's University Press, Montreal. Discontinued after Volume 16 (1975).

Bibliography on Cold Regions Science and Technology. Cold Regions Research and Engineering Laboratory (CRREL), Hanover, New Hampshire 03755, U.S.A. A continuing publication of the Library of Congress. The first 15 volumes were issued by the Snow, Ice and Permafrost Research Establishment (SIPRE). The first 19 volumes were entitled *Bibliography on Snow, Ice and Permafrost*. Volumes 20 through 22 were entitled *Bibliography on Snow, Ice and Frozen Ground*. Current title was adopted with Volume 23.

Biuletyn Peryglacjalny [*Periglacial Bulletin*]. Lodz, Poland.

Canadian Geotechnical Journal. National Research Council of Canada, Ottawa, Ontario K1A OR6, Canada.

CRREL Technical Publications. A publication of this title, which may be requested from the Cold Regions Research and Engineering Laboratory (CRREL), Hanover, New Hampshire 03755, U.S.A. lists CRREL reports of various types, draft translations, miscellaneous publications, and the published volumes of the *Bibliography on Cold Regions Science and Technology*. It also lists Arctic Construction and Frost Effects Laboratory (ACFEL)* technical reports, miscel-

*The Arctic Construction and Frost Effects Laboratory (ACFEL) and the Snow, Ice and Permafrost Establishment (SIPRE) were combined in 1961 to form the U.S. Army Cold Regions Research and Engineering Laboratory (CRREL).

laneous papers, and translations (including publications by the antecedent Frost Effects Laboratory and Permafrost Division organizations) and includes reports and translations of the former Snow, Ice and Permafrost Research Establishment (SIPRE).

DBR Publications. The Division of Building Research, National Research Council of Canada, Ottawa, Ontario K1A 0R6 Canada, has been a major world source of cold regions scientific and technical information since 1947. Supplements to the extensive *Division of Building Research List of Publications* are issued periodically.

Frost i Jord (Frost Action in Soils). Norwegian Committee on Permafrost, Oslo, Norway.

Meddelelser om Grønland. Reitzels Forlag, Copenhagen, Denmark.

Polar Record. Scott Polar Research Centre, Cambridge, England.

Problems of the North. National Research Council of Canada, Ottawa, Canada (trans. of *Problemy Severa,* Academy of Sciences, Moscow, U.S.S.R.).

Proceedings, Cold Regions Specialty Conference: Applied Techniques for Cold Environments. Held at Anchorage, Alaska, 17–19 May 1978, in 2 volumes, American Society of Civil Engineers, New York.

Proceedings, International Permafrost Conferences: *Permafrost International Conference,* Purdue University, November 1963, NAS-NRC Publication No. 1287, published 1966, NAS-NRC, Washington, D.C. *Second International Permafrost Conference*. Yakutsk, U.S.S.R., July 1973, 2 volumes (North American contribution and U.S.S.R. contribution), National Academy of Sciences, Washington, D.C. *Third International Permafrost Conference*. Edmonton, Alberta, Canada, July 1978, 2 volumes, National Research Council of Canada, Ottawa.

Quaternary Research. Academic Press, New York.

Reports, Institute of Polar Studies. Ohio State University, Columbus, Ohio.

The following publication is not devoted exclusively to the arctic or subarctic, but it does contain significant papers on permafrost or seasonal-frost-related topics of value to arctic and subarctic investigators:

Soil Mechanics and Foundation Engineering. A periodical translated from the Russian and published by Consultants Bureau, New York and London.

Sources for reference publications

U.S. Army Technical Manuals are distributed by:

U.S. Army AG Publications Center
1655 Woodson Rd.
St. Louis, MO 63114

U.S. Air Force Manuals are distributed by:

U.S. Air Force Publications
Distribution Center
2800 Eastern Blvd.
Baltimore, MD 21220

U.S. NAVFAC Manuals are distributed by:

U.S. Naval Publications and Forms Center
5801 Tabor Ave.
Philadelphia, PA 19120

U.S. Department of Defense Military Standards are distributed by:

U.S. Naval Publications and Forms Center
5801 Tabor Ave.
Philadelphia, PA 19120

CRREL, ACFEL and Waterways Experiment Station publications are available from: National Technical Information Service (NTIS) Springfield, Virginia 22151

References

For additional information on sources of technical literature, including addresses, see Appendix C.

Aitken, G. W. (1974). *Reduction in frost heave by surcharge load.* CRREL Technical Report 184.

——— (1965). *Ground temperature observations, Kotzebue, Alaska.* CRREL Technical Report 108.

Aldrich, H. P. and Paynter, H. M. (1953). *Analytical studies of freezing and thawing of soils.* ACFEL Technical Report 42.

American Association of State Highway and Transportation Officials (AASHTO) (1978). *Standard specifications for transportation materials and methods of sampling and testing, Part I. Specifications, Part II. Methods of sampling and testing.* Washington, D.C.

American Society for Testing Materials (ASTM) (1979). *Annual book of ASTM standards, Part 19. Soil and rock, building stones, peat.* ASTM, Philadelphia, Pennsylvania.

Arctic Construction and Frost Effects Laboratory (ACFEL) (1963). *Approach Roads, Greenland 1956–1957 Program.* ACFEL Technical Report 76, Waterways Experiment Station Technical Report 3-505, Report 2, Vicksburg, Mississippi.

——— (1958). *Cold room studies, third interim report of investigations.* ACFEL Technical Report 43.

——— (1954). *Depth of snow cover in the Northern Hemisphere.* Tables and color maps, folio size, ACFEL Technical Report 49.

——— (1954). *Groundwater studies, Fairbanks Permafrost Research Area.* CRREL Internal Report 40.

Babb, T. A. and Bliss, L. C. (1974). Susceptibility to environmental impact in the Queen Elizabeth Islands. *Arctic* **27,** 234–237.

Baranov, I. J. and Kudryavtsev, V. A. (1966). Permafrost in Eurasia. [First] *Permafrost international conference proceedings.* National Academy of Sciences–National Research Council Publication 1287, 98–102.

Barnes, D. F. (1966). Geophysical methods for delineating permafrost. [First] *Permafrost international conference proceedings.* National Academy of Sciences–National Research Council Publication 1287, 349–355.

Barnett, D. M.; Edlund, S. A.; and Hodgson, D. A. (1975). Sensitivity of surface materials and vegetation to disturbance in the Queen Elizabeth Islands: an approach and commentary. *Arctic* **28,** 74–76.

Bell, T. (1979). Change in the valley. *Alaska,* September 1979, 13–14.

Bentley, C. F. (1978). Canada's agricultural land resources and the world food problem. *International Society of Soil Science, 11th Congress* (Edmonton). Plenary papers, vol. 2, 1–26.

Berg, R. L.; Brown, J.; and Haugen, R. K. (1978). Thaw penetration and permafrost conditions associated with the Livengood to Prudhoe Bay Road, Alaska. In *Proceedings, third international conference on permafrost,* vol. 1, 615–621. National Research Council of Canada, Ottawa.

——— and Aitken, G. W. (1973). Some passive methods of controlling geocryological conditions in roadway construction. In *North American contribution, second international conference on permafrost,* National Academy of Sciences, Washington, D.C., 581–586.

Berson, G. Z.; Ivanovskii, S.R.; Saitburkhanov, S.R.; and Abzaev, I.A. (1968). [Vegetable growing in the far north and methods of increasing its efficiency.] *Problemy Severa,* Soviet Acad. Sci. **13:** 177–187 (National Research Council of Canada trans. *Problems of the North* **13:** 243–257).

Bilello, M. A. (1964). *Ice thickness observations in the North American arctic and subarctic.* CRREL Special Report 43/2.

——— (1961). *Ice thickness observations in the North American arctic and subarctic.* CRREL Special Report 43/1.

Bilello, M. A. and Bates, R. E. (1975). *Ice thickness observations in the North American arctic and subarctic.* CRREL Special Report 43/7.

——— and Bates, R. E. (1972). *Ice thickness observations in the North American arctic and subarctic.* CRREL Special Report 43/6.

——— and Bates, R. E. (1971). *Ice thickness observations in the North American arctic and subarctic.* CRREL Special Report 43/5.

——— and Bates, R. E. (1969). *Ice thickness observations in the North American arctic and subarctic.* CRREL Special Report 43/4.

——— and Bates, R. E. (1966). *Ice thickness observations in the North American arctic and subarctic.* CRREL Special Report 43/3.

Blake, W., Jr. (1974). Periglacial features and landscape evolution, Central Bathurst Island, District of Franklin. *Geol. Surv. Can.,* Pap. 74–1 (Part B), 235–244.

Bliss, L. C. (1978). Polar climates: their present agricultural uses and their estimated potential production in relation to soils and climate. *International Society of Soil Science, 11th Congress* (Edmonton). Plenary papers, vol. 2, 70–90.

Boughner, C. C. and Potter, J. C. (1953). Snow cover in Canada. *Weatherwise* **6,** 155–159, 170–171.

Brown, J. (1966). Ice-wedge chemistry and related frozen ground processes, Barrow, Alaska. [First] *Permafrost international conference proceedings.* National Academy of Sciences–National Research Council Publication 1287, 94–98.

Brown, R. J. E. and Kupsch, W. O. (1974). *Permafrost terminology.* NRCC 14274, National Research Council of Canada, Ottawa.

Brown, R. J. E. and Péwé, T. L. (1973). Distribution of permafrost in North America and its relationship to the environment: a review, 1963–1973. In *North American contribution to the second international conference on permafrost.* National Academy of Sciences, Washington, D.C., 71-100.

Bryson, R. A. and Hare, F. K. (1974). The climates of North America. In *World survey of climatology, vol. 11, Climates of North America* (R. A. Bryson and F. K. Hare, eds.). Elsevier Science Publishing, New York, I, 1–47.

Cailleux, A. and Taylor, G. (1954). *Cryopédologie, étude des sols gelés.* Hermann, Paris.

Casagrande, A. (1948). Classification and identification of soils. *Trans. Am. Soc. Civ. Eng.* **113,** 901–930 w/discussions.

——— (1931). Discussion on frost heaving. *Proc. Highway Res. Board* **11**(1), 168–172.

Clarke, G. R. (1957). *The study of the soil in the field.* Oxford University Press, Oxford.

Clayton, J. S.; Ehrlich, W. A.; Cann, D. B.; Day, J. H.; and Marshall, I. B. (1977). *Soils of Canada,* vol. 1. Printing and Publishing Supply and Services Canada, Ottawa.

Cold Regions Research and Engineering Laboratory (CRREL) (1962). *Ground temperature observations, Fort Yukon, Alaska.* CRREL Technical Report 100.

Crory, F. E. (1973). Settlement associated with the thawing of permafrost. In *North American contribution, second international conference on permafrost,* National Academy of Sciences, Washington, D.C., 599–607.

——— (1966). Pile foundations in permafrost. [First] *Permafrost international conference proceedings.* National Academy of Sciences–National Research Council Publication 1287, 467–472.

——— and Reed, R. E. (1965). *Measurement of frost heaving forces on piles.* CRREL Technical Report 145.

Czudek, T. and Demek, J. (1970). Thermokarst in Siberia and its influence on the development of lowland relief. *Quat. Res. N. Y.* **1,** 103–120.

Dempsey, B. J. and Thompson, M. R. (1969). *A heat transfer model for evaluating frost action and temperature related effects on multilayered pavement systems.* Professional Report IHR 401, University of Illinois.

Douglas, L. A. and Tedrow, J. C. F. (1960). Tundra soils of arctic Alaska. *Proceedings 7th International Congress of Soil Science* (Madison), vol. 4, Comm. V, 291–304.

Drew, J. V. (1957). *A pedologic study of Arctic Coastal Plain soils near Point Barrow, Alaska.* Unpublished Ph.D. thesis, Rutgers University.

——— and Tedrow, J. C. F. (1962). Arctic soil classification and patterned ground. *Arctic* **15,** 109–116.

——— Tedrow, J. C. F.; Shanks, R. E.; and Koranda, J. J. (1958). Rate and depth of thaw in arctic soils. *Am. Geophys. Union Trans.* **39,** 279–295.

Ecology of the Subarctic regions. (1970). UNESCO Helsinki Symp., UNESCO, Paris.

Fedoroff, N. (1966). Les cryosols. *Sci. Sol.* **2,** 77–110.

Ferrians, O. J. and Hobson, G. D. (1973). Mapping and predicting permafrost in North America: a review, 1963–1973. In *North American contribution, second international conference on permafrost,* National Academy of Sciences, Washington, D.C., 479–498.

Fridland, V. M. (1976). *Pattern of the soil cover.* Keter Pub. House, Jerusalem. (Translated from the Russian. Original published in 1972 by the Soviet Academy of Sciences, Moscow.)

Frost, R. E. (1950). *Evaluation of soils and permafrost conditions in the territory of Alaska by means of aerial photographs,* 2 vols. Prepared by Engineering Experiment Station, Purdue University, for St. Paul Dist., Corps of Engineers. Technical Report 34 in ACFEL list of reports.

——— McLerran, J. H.; and Leighty, R. D. (1966). Photointerpretation in the Arctic and Subarctic. [First] *Permafrost international conference proceedings.* National Academy of Sciences–National Research Council Publication 1287, 343–348.

Frost Effects Laboratory (1952). *Investigation of description, classification and strength properties of frozen soils, FY 1951.* ACFEL Technical Report 40/SIPRE Report 8, in 2 vols.

——— (1950). *Preparations of frost effects laboratory for Project Overheat.* ACFEL Technical Report TR 27.

——— (1947). *Report of investigations, investigation of construction and maintenance of airdromes on ice, 1946–1947.* ACFEL Technical Report 8.

Fulwider, C. W. and Aitken, G. W. (1962). Effect on surface color on thaw penetration beneath a pavement in the Arctic. In *Proceedings first international conference on the structural design of asphalt pavements.* Univ. of Michigan, Ann Arbor, Michigan, 605–610.

Garg, O. P. (1973). In situ physiocomechanical properties of permafrost using geophysical techniques. In *North American contribution, second international conference on permafrost.* National Academy of Sciences, Washington, D.C., 508–517.

Gasser, G. W. (1951). A brief account of agriculture in Alaska. *Science in Alaska*. Proceedings second Alaskan scientific conference, Mt. McKinley National Park, Alaska, 86–92.

George, W. (1973). Analysis of the proposed Little Chena River, earthfilled nonretention dam, Fairbanks, Alaska. In *North American contribution, second international conference on permafrost*. National Academy of Sciences, Washington, D.C., 636–648.

Gersper, P. L. and Challinor, J. L. (1975). Vehicle perturbation effects upon a tundra soil-plant system: I. Effects on morphological and physical environmental properties of the soils. *Soil Sci. Soc. Am. Proc.* **39,** 737–744.

Gold, L. W. (1971). Use of ice covers for transportation. *Can. Geotech. J.* **8,** 170.

——— and Lachenbruch, A. H. (1973). Thermal conditions in permafrost—a review of North American literature. In *North American contribution, second international conference on permafrost*. National Academy of Sciences, Washington, D.C., 3–25.

Gow, A. J. and Williamson, T. (1976). *Rheological implications of the internal structure and crystal fabrics of the West Antarctic ice sheet as revealed by deep core drilling at Byrd Station*. CRREL Report 76-35.

Granberg, H. B. (1973). Indirect mapping of the snowcover for permafrost prediction at Schefferville, Quebec. In *North American contribution, second international conference on permafrost*. National Academy of Sciences, Washington, D.C., 113–120.

Gravé, N. A. (1968). Merzlye Tolshchi Zemli [The Earth's Permafrost Beds]. *Priroda* **1,** 46–53 (trans. Defence Res. Bd., Canada, July 1968, T499R).

Hamelin, L. E. and Cook, F. A. (1967). *Illustrated glossary of periglacial phenomena*. University of Laval Press, Quebec.

Handy, R. L. and Fenton, T. E. (1977). Particle size and mineralogy in soil taxonomy. In *Soil Taxonomy and soil properties*. Transportation Research Record 642. Transportation Research Board, Washington, D.C., 13–19.

Hoekstra, P.; Chamberlain, E.; and Frate, A. (1965). *Frost heaving pressures*. CRREL Research Report 176.

Hoppe, G. (1959). Glacial morphology and inland ice recession in northern Sweden. *Geogr. Ann.* (Stockholm) **4,** 193–244.

Hunter, J. A. M. (1973). The application of shallow seismic methods to mapping of frozen surficial materials. In *North American contribution, second international conference on permafrost*. National Academy of Sciences, Washington, D.C. 527–535.

Hustich, I. (1966). On the forest-tundra and the northern tree-lines. *Ann. Univ. Turku (Kevo)*, A, II: 36 (41 pp.).

——— (1953). The boreal limits of conifers. *Arctic* **6,** 149–162.

Hvorslev, M. J. (1949). *Subsurface exploration and sampling of soils for engineering purposes*. U.S. Army Waterways Experiment Station, Vicksburg, Mississippi.

——— and Goode, T. B. (1966). Core drilling in frozen soils. In [First] *Permafrost international conference proceedings*. National Academy of Sciences–National Research Council Publication 1287, 364–377.

——— (1957). Core drilling in frozen ground. *Seventh annual drilling symposium—exploration drilling,* University of Minnesota, Minneapolis, 114–126.

Ivanova, E. N. (1956). Sistematika pochv severnoi chasti Evrpeiskoi territorii SSSR [Classification of soils of the northern parts of the European U.S.S.R.]. *Pochvovedenie* **1,** 70–88 (OTS 61–11496).

Ivanovskii, A. I. (1963). [Transformation of nature and ways of developing agriculture in the far north.] *Problemy Severa*. Soviet Academy of Science **7,** 5–21 (National Research Council of Canada, Trans. *Problems of the North* **7,** 1–19).

Jahn, A. (1971). *Zagadnienia strefy peryglacjalnej [Problems of the periglacial zone]*. Polish Scientific Publishers, Warsaw. (U.S. Dept. Comm. TT72-54011.)

——— (1961). [Quantitative analysis of some periglacial processes in Spitsbergen.] Wroclawski im. Boleslawa Bieruta Zeszyty Naukowe, *Nauki Przyrodnicze,* Ser. B5 (Nauka o ziemi II), 1–34.

Johnson, L. and Van Cleve, K. (1976). *Revegetation in arctic and subarctic North America —a literature review.* CRREL Technical Report 76-15.

Johnston, G. H. (1966) *Engineering site investigations in permafrost areas.* National Academy of Sciences–National Research Council Publication 1287, 371–374.

Kallio, A. and Rieger, S. (1969). Recession of permafrost in a cultivated soil of interior Alaska. *Soil Sci. Soc. Am. Proc.* **33,** 430–432.

Kaplar, C. W. (1974). *Freezing test for evaluating relative frost susceptibility of various soils.* CRREL Technical Report 250.

——— (1965). *Some experiments to measure frost heaving force in a silt.* CRREL Technical Note. (Unnumbered)

——— (1954). *Investigation of the strength properties of frozen soils, FY 1953.* ACFEL Technical Report 48, 2 vols.

——— (1953). *Investigation of the strength properties of frozen soils, FY 1952.* ACFEL Technical Report 44, 2 vols.

Kersten, M. S. (1949). *Laboratory research for the determination of the thermal properties of soils.* Report by Engineering Experiment Station, University of Minnesota to St. Paul District, U.S. Army Corps of Engineers, ACFEL Technical Report TR-23. Also *The thermal properties of soils.* Bull. 28. University of Minnesota Engineering Experiment Station.

Keune, R. and Hoekstra, P. (1967). *Calculating the amount of unfrozen water from moisture characteristic curves.* CRREL Special Report 114.

Kitze, F. F. (1956). *Some experiments in drive sampling in frozen ground.* ACFEL Miscellaneous Paper 16.

Korol, N. (1955). Agriculture in the zone of perpetual frost. *Science* **122,** 680–682.

Kritz, M. A. and Wechsler, A. E. (1967). *Surface characteristics, effect on thermal regime, Phase II.* CRREL Technical Report 189.

Kubiëna, W. L. (1970). *Micromorphological features of soil geography.* Rutgers University Press, New Brunswick, N.J.

Lachenbruch, A. H. (1970). *Some estimates of the thermal effects of a heated pipeline in permafrost.* (U.S.) Geological Survey Circular 632.

——— (1962). *Mechanics of thermal contraction cracks and ice-wedge polygons in permafrost.* Geological Society of America, Special Paper 70.

——— and Marshall, B. V. (1969). Heat flow in the arctic. *Arctic* **22,** 300–311.

Lagutin, B. L., ed. (1946). *Data on the problem of ice crossings. Translations, appendix A to report of investigations, investigation of construction and maintenance of airdromes on ice, fiscal year 1950.* ACFEL Technical Report 29.

Lambe, T. W. (1951). *Soil testing for engineers.* Wiley, New York.

Lane, K. S. (1948). Treatment of frost sloughing slopes. *Proceedings, second international conference on soil mechanics and foundation engineering,* Rotterdam, 201–203.

Lange, G. R. (1973a). Investigation of sampling perennially frozen alluvial gravel. In *North American contribution, second international conference on permafrost.* National Academy of Sciences, Washington, D.C., 535–541.

——— (1973b). *An investigation of core drilling in perennially frozen gravels and rock.* CRREL Technical Report 245.

——— (1968). *Rotary drilling and coring in permafrost.* CRREL Technical Report 95.

Lawson, D. E.; Brown, J.; Everett, K. R.; Johnson, A. W.; Komárková, V.; Murray, B. M.; Murray, D. F.; and Webber, P. J. (1978). *Tundra disturbances and recovery following the 1949 exploratory drilling, Fish Creek, northern Alaska.* CRREL Report 78-28.

Lee, H. A. (1962). Method of deglaciation, age of submergence, and rate of uplift west and east of Hudson Bay, Canada. *Biul. Periglacjalny* (Lodz) **11,** 239–245.

Linell, K. A. (1973a). Risk of uncontrolled flow from wells through permafrost. In *North American contribution, second international conference on permafrost.* National Academy of Sciences, Washington, D.C., 462–468.

——— (1973b). Long-term effects of vegetative cover on permafrost stability in an area of discontinuous permafrost. In *North American contribution second international conference on permafrost.* National Academy of Sciences, Washington, D.C., 688–693.

——— (1960). Frost action and permafrost. In *Highway engineering handbook,* Section 13. K. B. Woods (ed.), McGraw-Hill, New York.

——— (1957). Airfields on permafrost. *Proc. Am. Soc. Civ. Eng., J. Air Transport Div.* **83**(1), 1316-1–1326-15.

——— and Haley, J. F. (1952). Investigation of the effect of frost action on pavement supporting capacity. In *Frost action in soils, a symposium.* Highway Research Board Special Report No. 2. Highway Research Board, National Academy of Sciences, Washington, D.C., 295–325.

——— Hennion, F. B.; and Lobacz, E. F. (1963). Corps of Engineers pavement design in areas of seasonal frost. In *Pavement design in frost areas, II. Design considerations.* Highway Research Record No. 33. Highway Research Board, National Academy of Sciences, Washington, D.C., 76–136.

——— and Johnston, G. H. (1973). Engineering design and construction in permafrost regions: a review. In *North American contribution, second international conference on permafrost.* National Academy of Sciences, Washington, D.C., 553–575.

——— and Kaplar, C. W. (1966). Description and classification of frozen soils. [First] *Permafrost international conference proceedings.* National Academy of Sciences–National Research Council Publication 1287, 481–487. Also CRREL Technical Report 150.

——— and Kaplar, C. W. (1959). The factor of soil and material type in frost action. In *Highway pavement design in frost area, a symposium: part 1. Basic considerations.* Highway Research Board Bulletin 225. Highway Research Board. National Academy of Sciences–National Research Council, Washington, D.C., 81–128.

——— Lobacz, E. F.; et al (1980). *Design and construction of foundations in areas of deep seasonal frost and permafrost.* Includes material contributed by W. F. Quinn, F. H. Sayles, H. W. Stevens, G. W. Aitken, F. E. Crory, C. W. Fulwider, F. J. Sanger, and W. N. Tobiasson. CRREL Special Report 80–34.

Lobacz, E. F. and Quinn, W. F. (1966). Thermal regime beneath buildings constructed on permafrost. [First] *Permafrost international conference proceedings.* National Academy of Sciences–National Research Council Publication 1287, 247–252.

Lunardini, V. J. (1978). Theory of n-factors and correlation of data. In *Proceedings of the third international conference on permafrost,* vol. 1. National Research Council of Canada, Ottawa, 40–46.

Makeev, O. V. (1978). Optimum soil utilization systems under differing climatic restraints: polar regions. *International Society of Soil Science, 11th Congress* (Edmonton). Plenary papers, vol. 2, 27–69.

McCoy, J. E. (1965). *Excavations in frozen ground Alaska, 1960–61.* CRREL Technical Report 120.

Meinardus, W. (1930). Zum Jahreszeitlichen Gang der Beleuchtung in den Polargebieten. *Arktis* **3,** 4–6.

Mellor, M. (1975). *Cutting frozen ground with disc saws.* CRREL Technical Report 261.

——— (1965). *Blowing snow.* CRREL Monograph III-A3c.

Middendorf, A. von. (1864). Übersicht über die Natur Nord und Ostsibiriens: 4. Die Gewächse Sibiriens. *Sibirische Reisen* (St. Petersburg) **4**(1).

Mollard, J. D. and Pihlainen, J. A. (1966). Airphoto interpretation applied to road selection in the Arctic. [First] *Permafrost international conference proceedings.* National Academy of Sciences–National Research Council Publication 1287, 381–387.

Nixon, J. F. and Morgenstern, N. R. (1973). Practical extensions to a theory of consolidation for thawing soils. In *North American contribution, second international conference on permafrost.* National Academy of Sciences, Washington, D.C. 369–377.

Penner, E. and Ueda, T. (1978). A soil frost susceptibility test and a basis for interpreting heaving rates. In *Proceedings of the third international conference on permafrost,* vol. 1. National Research Council of Canada, Ottawa, 721–727.

Péwé, T. L. (1954). Effect of permafrost on cultivated fields, Fairbanks area, Alaska. *U.S. Geol. Surv. Bull.* 989-F.

——— and Paige, R. A. (1963). Frost heaving of piles with an example from Fairbanks, Alaska. *U.S. Geol. Surv. Bull.* 1111-1.

Pihlainen, J. A. (1965). *Construction in muskeg.* CRREL Technical Report 134.

——— and Johnston, G. H. (1963). *Guide to a field description of permafrost for engineering purposes.* Technical Memo 79, NRC 7576, Assoc. Committee on Soil and Snow Mechanics, National Research Council of Canada, Ottawa.

Polunin, N. (1951). The real arctic: suggestions for its delineation, subdivision and characterization. *J. Ecol.* **39,** 303–315.

Radd, F. J. and Oertle, D. H. (1973). Experimental pressure studies of frost heave mechanisms and the growth-fusion behavior of ice. In *North American contribution, second international conference on permafrost.* National Academy of Sciences, Washington, D.C., 377–384.

Radforth, N. W. (1952). Suggested classification of muskeg for the engineer. *Eng. J.* **35**(11), 1199–1210.

——— and Brawner, C. O. (eds.). (1977). *Muskeg and the northern environment in Canada.* University of Toronto Press, Toronto.

Roethlisberger, H. (1972). *Seismic exploration in cold regions.* CRREL Monograph II-A2a.

Roggensack, W. D. and Morgenstern, N. R. (1978). Direct shear tests on natural fine-grained permafrost soils. In *Proceedings of the third international conference on permafrost,* vol. 1, National Research Council of Canada, Ottawa, 728–735.

Ryder, T. (1954). *Compilation and study of ice thickness in the Northern Hemisphere.* ACFEL Technical Report TR 47.

——— (1953). *Compilation and study of ice thicknesses in the Northern Hemisphere. 1952–1953. Tabulations of ice thickness data.* ACFEL Technical Report TR 47/A.

Sachs, V. N. and Strelkov, S. A. (1961). Mesozoic and Cenozoic of the Soviet arctic. *Geology of the arctic* (G. O. Raasch, ed.). University of Toronto Press, Toronto, 1, 48–67.

Sanderson, M. (1948). Drought in the Canadian Northwest. *Geogr. Rev.* **38,** 289–299.

Sanger, F. J. (1968). *Ground freezing in construction. Proc. Am. Soc. Civ. Eng.,* **94** (SM1). Proc. paper 5743.

——— (1966). Degree-days and heat conduction in soils. In [First] *Permafrost international conference proceedings.* National Academy of Sciences–National Research Council Publication 1287, 253–262.

Sayles, F. H. (1973). Triaxial and creep tests on frozen Ottawa sand. In *North American contribution, second international conference on permafrost.* National Academy of Sciences, Washington, D.C., 384–391.

——— (1968). *Creep of frozen sands.* CRREL Technical Report TR-190.

Scott, W. J.; Sellmann, P. V.; and Hunter, J. A. M. (1979). Geophysics in the study of permafrost. In *Proceedings of the third international conference on permafrost* (Edmonton). vol. 2. National Research Council of Canada, Ottawa, 93–115.

Sebastyan, C. Y. (1966). Preliminary site investigation for the foundation of structures and pavements in permafrost. In [First] *Permafrost international conference proceedings.* National Academy of Sciences–National Research Council Publication 1287, 387–394.

Sellmann, P. V. and Brown, J. (1965). *Coring of frozen ground, Barrow, Alaska.* CRREL Special Report 81.

Sjors, H. (1961). Surface patterns of boreal peatland. *Endeavour* **20,** 217–224.

Slavin, S. V. (1958). [On the concept "Soviet North"]. *Problemy Severa* Soviet Acad. Sci. **2,** 253–265 (National Research Council of Canada trans. *Problems of the North* **7,** 1–19).

Soil Survey Staff (1975). *Soil taxonomy.* U.S. Dept. Agriculture Handbook 436. U.S. Government Printing Office, Washington, D.C.

——— (1951). *Soil survey manual.* Agriculture Handbook 18. U.S. Government Printing Office, Washington, D.C.

Spangler, M. G. (1960). Engineering characteristics of soil and soil testing. In *Highway engineering handbook* (K. B. Woods, ed.). McGraw-Hill, New York, 8-3–8-63.

Stearns, S. R. (1965). *Selected aspects of geology and physiography of the cold regions.* CRREL monograph 1-A1.

——— (1966). *Permafrost (perennially frozen ground).* CRREL Monograph I-A2.

Targulian, V. O. (1971). *Pochvoobrazovanie i vyvetrivanie v kholodnykh gumidnykh oblastiakh* [*Soil formation and weathering in cold humid regions*]. Soviet Academy of Science, Moscow.

Tedrow, J. C. F. (1977). *Soils of the polar landscapes.* Rutgers University Press, New Brunswick, N.J.

——— (1970). Soils of the subarctic regions. *Ecology of the subarctic regions.* Proceedings UNESCO Helsinki Symposium, UNESCO, Paris, 189–205.

——— (1966). Polar desert soils. *Soil Sci. Soc. Am. Proc.* **30,** 381–387.

——— and Cantlon, J. E. (1958). Concepts of soil formation and classification in arctic regions. *Arctic* 11, 166–179.

——— Drew, J. V.; Hill, D. E.; and Douglas, L. A. (1958). Major genetic soils of the Arctic Slope of Alaska *J. Soil Sci.* 9, 33–45.

——— and Hill, D. E. (1955). Arctic brown soil. *Soil Sci.* **80,** 265–275.

Terzaghi, K. (1952). Permafrost. *J. Boston Soc. Civ. Eng.* **39**(1), 1–50.

——— and Peck, R. B. (1967). *Soil mechanics in engineering practice* (2nd ed.). Wiley, New York.

Thomas, M. K. (1953). *Climatological atlas of Canada.* NRC 3151, Canadian Meteorological Branch and Div. of Building Research, National Research Council of Canada, Ottawa.

Thompson, M. R. and Robnett, Q. L. (1979). Resilient properties of subgrade soils. *Transp. Eng. J., Proc. Am. Soc. Civ. Eng.* 105 (TE1), 71–89.

Tikhomirov, B. A. (1962). The treelessness of the tundra. *Polar Rec.* **11**(70), 24–30.

Troll, C. (1958). *Structure soils, solifluction, and frost climates of the earth.* Snow, Ice and Permafrost Research Establishment (Wilmette, Ill.) trans. 43. [Originally published in *Geol. Rundschau* (Leipzig) **34,** 545–694. 1943.]

Tyurin, I. V.; Gerasimov, I. P.; Ivanova, E. N.; and Nosin, V. A. (eds.) (1959). *Pochvennaya s ëmka*. Akademiya Nauk SSSR. [Soil Survey, Academy of Sciences, U.S.S.R.] Trans. by Israel Program for Sci. Trans. IPST No. 1356. U.S. Dept. Commerce, Washington, D.C.

U.S. Army (1963). *Terrain evaluation in Arctic and Subarctic regions*. Technical Manual TM 5-852-8.

——— (1962). *Arctic construction*. Technical Manual TM 5-349.

U.S. Army Corps of Engineers (1970). *Laboratory soils testing*. Engineer Manual EM 1110-2-1906. Soil tests for civil works projects.

——— (1953). *The Unified Soil Classification System*. U.S. Army Waterways Experiment Station Technical Memo 3-357, rev. 1957 and 1960, with Appendixes A and B, Vicksburg, Mississippi.

U.S. Army/U.S. Air Force (1966a). *Arctic and Subarctic construction, general provisions*. Technical Manual TM 5-852-1/AFM 88-19, chap. 1.

——— (1966b). *Site selection and development, Arctic and Subarctic construction*. Technical Manual TM 5-852-2/AFM 88-19, chap. 2.

——— (1966c). *Calculation methods for determination of depths of freeze and thaw in soils, Arctic and Subarctic construction*. Technical Manual TM 5-852-6/AFM 88-19, chap. 6.

——— (1966d). *Pavement design for frost conditions*. Technical Manual TM 5-818-2/AFM 88-6, chap. 4.

——— (1961). *Procedures for foundation design of buildings and other structures (except hydraulic structures)*. Technical Manual TM 5-818-1/AFM 88-3, chap. 7.

——— (forthcoming). *Foundations for structures, Arctic and Subarctic construction*. Technical Manual TM 5-852-4/AFM 88-19, chap. 4.

U.S. Bureau of Reclamation (1963). *Earth manual* (1st ed. rev.). Washington, D.C.

U.S. Dept. of Defense (rev. 1968). *Unified classification system for roads, airfields, embankments and foundations*. Military Standard MIL-STD-619B, Washington, D.C.

U.S. Federal Housing Administration (1959). *Engineering soil classification for residential developments*. U.S. Government Printing Office, Washington, D.C.

U.S. Naval Facilities Engineering Command (1975). *Design Manual. Cold Regions Engineering*. NAVFAC DM-9, Alexandria, Virginia.

U.S. Naval Observatory (annual). *The American ephemeris and nautical almanac*. U.S. Government Printing Office, Washington, D.C. Also published by the Royal Greenwich Observatory as *The Astronomical Ephemeris*, Her Majesty's Stationery Office, 49 High Holborn, W.C. 1.

U.S. Navy Weather Research Facility (1962). *Arctic Forecast Guide*. NWRF 16-0462-058. Prepared by Dr. R. J. Reed, Norfolk, VA.

Washburn, A. L. (1973). *Periglacial processes and environments*. Arnold, London.

——— (1967). Instrumental observations of mass wasting in the Mesters Vig District northeast Greenland. *Medd. om Grønland* **166**(4), 318.

——— (1956). Classification of patterned ground and review of suggested origins. *Geol. Soc. Am. Bull.* **67**, 823–865.

Wechsler, A. E. and Glaser, P. E. (1966). *Surface characteristics, effect on thermal regime, phase I*. CRREL Special Report SR-88.

Weyer, E. J., Jr. (1943). Day and night in the arctic. *Geogr. Rev.* **33**, 474–478.

Wilson, C. (1969). *Climatology of the cold regions: Northern Hemisphere II*. CRREL Monograph I-A3b.

——— (1967). *Climatology of the cold regions: introduction, Northern Hemisphere I*. CRREL Monograph I-A3a.

Partial list of conversion factors

AMERICAN CUSTOMARY AND METRIC TO SI

AREA	1 $in^2 = 6.4516 \times 10^{-4}\ m^2$
	1 $ft^2 = 0.0929\ m^2$
DENSITY	1 $lb/in^3 = 2.768 \times 10^4\ kg/m^3$
	1 $lb/ft^3 = 16.018\ kg/m^3$
	1 $g/cm^3 = 1 \times 10^3\ kg/m^3$
ENERGY	1 Btu = 1.055 kJ
	1 cal = 4.1868 J
	1 Btu/lb = 2.326 kJ/kg
	1 cal/g = 4.1868 kJ/kg
	1 $Btu/ft^3 = 37.259\ kJ/m^3$
	1 $cal/cm^3 = 4.1868\ MJ/m^3$
FLOW	1 gpm $= 6.309 \times 10^{-5}\ m^3/s$
FORCE	1 lbf = 4.4482 N
	1 kip $= 4.4482 \times 10^3$ N
HEAT FLUX	1 $Btu/ft^2/s = 11.356\ kW/m^2$
	1 $cal/cm^2/s = 4.1868 \times 10^4\ W/m^2$
	1 $kgcal/m^2/h = 1.163\ W/m^2$
LENGTH	1 in = 0.0254 m
	1 ft = 0.3048 m
	1 yd = 0.9144 m
	1 statute mile = 1.60935 km
MASS	1 lb = 0.45359 kg
POWER	1 J/s = 1W
	1 cal/s = 4.1868 W
PRESSURE	1 psi = 6.8948 kN/m^2
	1 psf = 0.04788 kN/m^2

TEMPERATURE	1°F = 5/9° C = 5/9 K
	$^{t}C = 5/9(t_F - 32) = t_K - 273.15$
	1 degree-day F = 5/9 degree-day C (or K)
THERMAL CONDUCTIVITY	1 Btu/ft h°F = 1.7307 W/mK
	1 Btu/in h ft^{2}° F = 0.1442 W/mK
	1 cal/cm s°C = 418.68 W/mK
THERMAL DIFFUSIVITY	1 ft^{2}/h = 2.5806 x 10^{-5} m^{2}/s
VELOCITY	1 mph = 0.44704 m/s
	1 ft/min = 5.080 x 10^{-3} m/s
VOLUME	1 m^{3} = 1.6387 x 10^{-5} m^{3}
	1 ft^{3} = 0.02832 m^{3}

Abbreviations of units

Btu = British thermal unit
cal = calorie
C = Celsius
cm = centimeter
F = Fahrenheit
ft = foot
f = force
g = gram
gpm = gallon per minute
h = hour
in = inch
J = joule
K = Kelvin
k = kilo
kg = kilogram
kip = 1000 pounds
lb = pound
m = meter
min = minute
mph = mile per hour
N = newton
psi = pound per square inch
psf = pound per square foot
s = second
t = temperature
W = watt
yd = yard

Author index

(Names mentioned in captions to figures and tables are included in this index.)

Subject index

(Page numbers in italics refer to related figures and tables.)

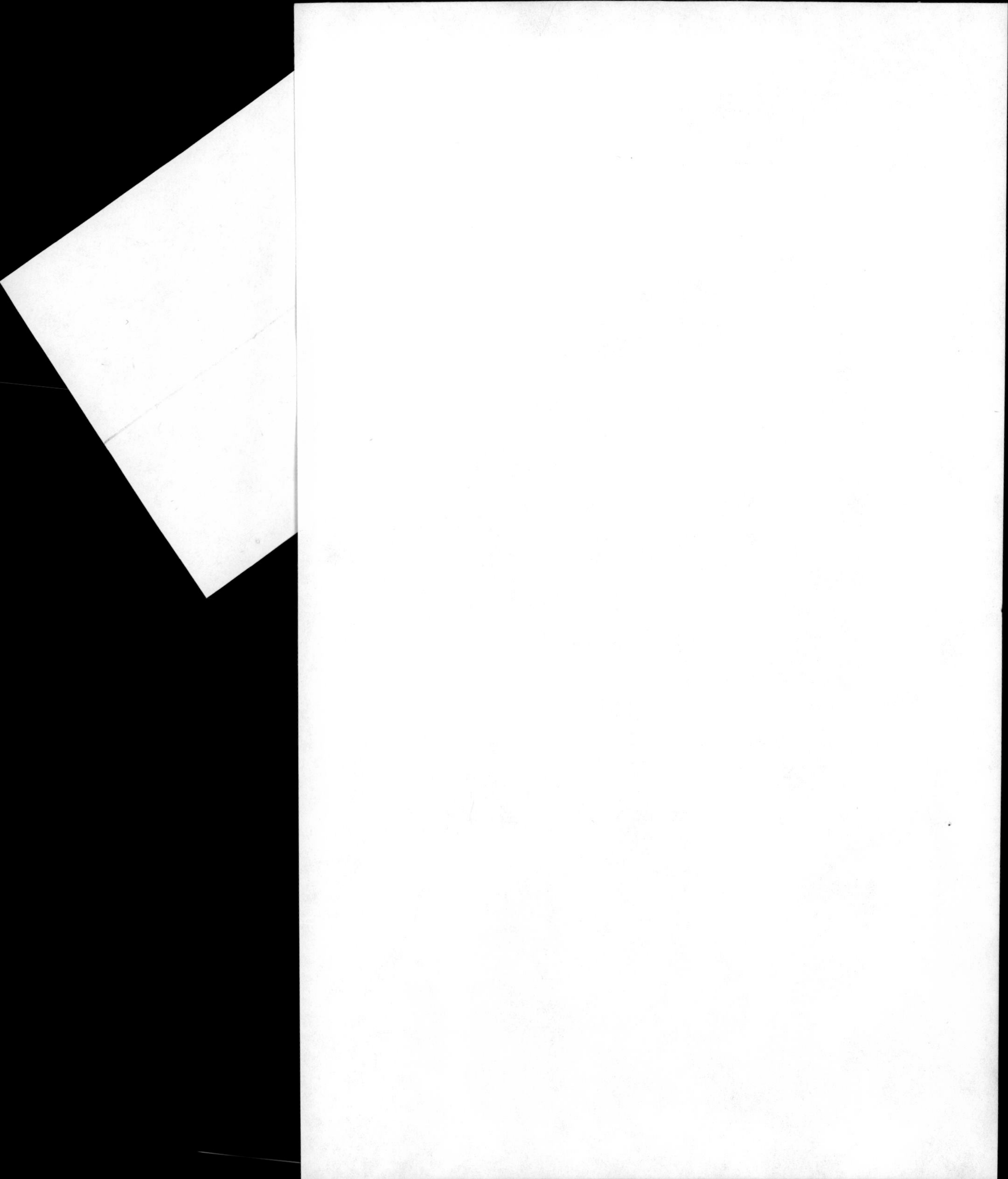

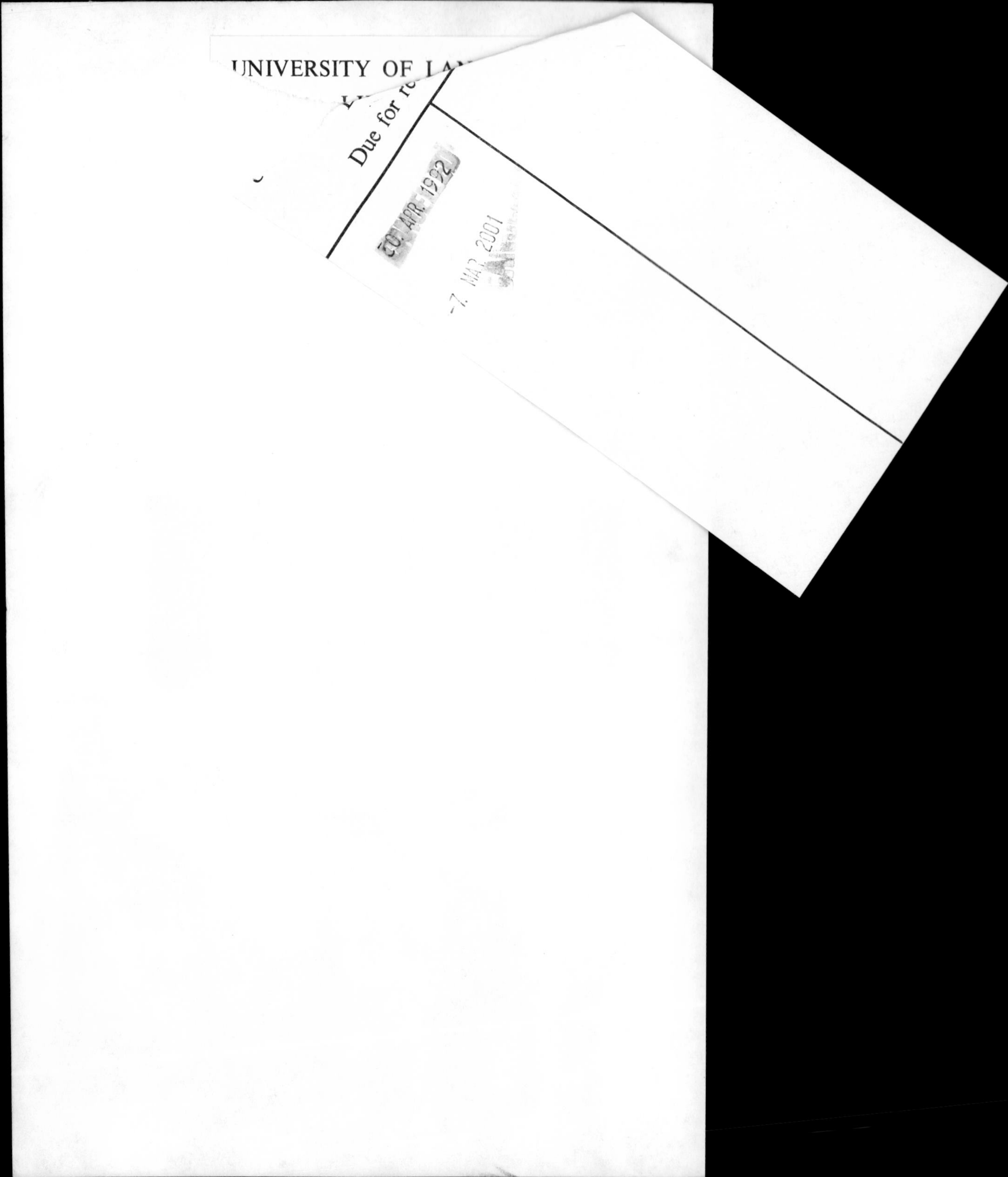
UNIVERSITY OF LA
Due for r
20. APR. 1992
-7. MAR. 2001